T0136061

About Island Press

Since 1984, the nonprofit Island Press has been stimulating, shaping, and communicating the ideas that are essential for solving environmental problems worldwide. With more than 800 titles in print and some 40 new releases each year, we are the nation's leading publisher on environmental issues. We identify innovative thinkers and emerging trends in the environmental field. We work with world-renowned experts and authors to develop cross-disciplinary solutions to environmental challenges.

Island Press designs and implements coordinated book publication campaigns in order to communicate our critical messages in print, in person, and online using the latest technologies, programs, and the media. Our goal: to reach targeted audiences—scientists, policymakers, environmental advocates, the media, and concerned citizens—who can and will take action to protect the plants and animals that enrich our world, the ecosystems we need to survive, the water we drink, and the air we breathe.

Island Press gratefully acknowledges the support of its work by the Agua Fund, Inc., The Margaret A. Cargill Foundation, Betsy and Jesse Fink Foundation, The William and Flora Hewlett Foundation, The Kresge Foundation, The Forrest and Frances Lattner Foundation, The Andrew W. Mellon Foundation, The Curtis and Edith Munson Foundation, The Overbrook Foundation, The David and Lucile Packard Foundation, The Summit Foundation, Trust for Architectural Easements, The Winslow Foundation, and other generous donors.

The opinions expressed in this book are those of the author(s) and do not necessarily reflect the views of our donors.

The Restoring Ecological Health to Your Land Workbook

Steven I. Apfelbaum and Alan Haney

With illustrations by Kirsten R. Vinyeta

ISLANDPRESS
Washington | Covelo | London

Library of Congress Cataloging-in-Publication Data

Apfelbaum, Steven I., 1954–
  The restoring ecological health to your land workbook / Steven I. Apfelbaum and Alan Haney.
     p. cm.
  Includes bibliographical references and index.
  ISBN-13: 978-1-59726-804-2 (pbk. : alk. paper)
  ISBN-10: 1-59726-804-6 (pbk. : alk. paper) 1. Restoration ecology. Apfelbaum, Steven I., Haney, Alan, II. Title.
  HQ541.15.R45A64 2012
  333.71′53—dc23

2011031269

Printed on recycled, acid-free paper ✪

Manufactured in the United States of America
10 9 8 7 6 5 4 3 2 1

Keywords: Ecological restoration, land management, adaptive management, ecological resilience, restoration practitioner, invasive species, mapping, soil survey, landscape change, restoration plan, ecological monitoring, financing restoration, budgeting restoration, ecosystem restoration, ecosystem monitoring

CONTENTS

It is in man's heart that the life of nature's spectacle exists; to see it, one must feel it.

Jean-Jacques Rousseau

This workbook provides detailed information for preparing and implementing an ecological restoration project. It closely follows our previous book, *Restoring Ecological Health to Your Land*[1], in which we describe an approach that has been consistently successful in restoring ecosystems. The methodology we describe is pragmatic. It is the result of years of experience with restoration of hundreds of ecosystems, including prairies, forests, deserts, wetlands, and streams around the world. It is our aim in this book to provide the basic tools and techniques, with enough instruction so that you can tailor them to the specific and unique aspects of your project.

We have organized this book in the sequence that should be followed for every project, beginning with evaluation of the land. We provide guidance for each step and sources for obtaining the essential information required to prepare a good restoration plan; then we progress to implementation of the plan. Monitoring results of restoration treatments and good record keeping are necessary to adaptively manage and refine restorations during implementation; we provide guidance for both. We also have included information on such things as financing and contracting for services and strategies for reducing costs.

This logical flow of preparing and implementing a restoration plan is important for three reasons: it facilitates communication with others, it supports your need to optimally utilize your investments of time and money, and it guides you in building the essential knowledge and understanding of the ecosystems you are restoring. Do not underestimate the importance of communications. If there is a most important lesson learned by the authors, it is the importance of developing a simple, understandable way to share the restoration project with the stakeholders, be they your own family, neighbors, financial partners, or volunteers. We emphasize careful organization and good communication throughout the book, primarily because we have seen good projects fail for lack of attention to these elements more than for any other reason.

Getting started is the hardest part of any difficult task. Restoration and management of ecosystems is complex, with a staggering array of things to consider. It is easy to get discouraged, and even the most experienced of us often feel overwhelmed. We have tried to identify the common barriers and stumbling points, and provide you with an approach for dealing with them that increases your comfort and confidence. The approach always should begin with learning your land, how it has changed, how it has been impaired, and why and where it is presently being stressed. Starting work on the land prematurely without this knowledge, or without the understanding of how to address the stressors, can cost you time and money, at best, or increase the problems on the land, at worst.

Finally, we offer something unique for our readers. To help ensure that you are successful, we have established a website (www.restoringyourland.com) and offer a one-hour, free consultation that will help you into and through the restoration process. Additional information about the website, such as how to apply for assistance, is found in appendix 4.

This book is part of the series The Science and Practice of Ecological Restoration, sponsored by the Society for Ecological Restoration and published by Island Press. Other books in the series are listed at the end of the book. Many are written by restoration experts for those experienced with at least some aspects of restoration ecology and will provide you with useful references for specific ecosystems like those you are planning to restore. These books contain details not included in this book; as you gain experience and knowledge about your restoration, this information will become more useful. The book you are holding in your hands, along with our previous book, differs from others in the series, because we have assumed as little as possible about your experience or formal training. If you are just getting into ecological restoration, we hope that our books will give you the information you need to begin.

In preparing both books, we have drawn from our personal experiences in restoring our own lands, as well as from hundreds of other projects with which we have been involved. We also drew from the experiences of others and applied our accumulated knowledge to lay out an approach to ecological restoration that we have found to be consistently successful.

We think of those engaged in ecological restoration as the "land-connection community." This community includes private landowners interested in conservation, land stewards working for land trusts or conservancies (one of the fastest growing conservation movements in the world), governmental and nongovernmental personnel engaged in ecosystem restoration and management, personnel working with for-profit corporations and developers, and volunteers. The common thread for all of us engaged in ecological restoration and ecosystem management is the ecological health of the land.

What Is the Meaning of "Ecological Health"?

Aldo Leopold explored ecological processes on his down-trodden central Wisconsin farm in his seminal book, *A Sand County Almanac*.[2] He introduced the concept of land health and related it to environmental ethics:

> A land ethic . . . reflects a conviction of individual responsibility for health of the land. Health is the capacity of the land for self-renewal. Conservation is our effort to understand and preserve this capacity. (Leopold 1949)

Nina Leopold Bradley, daughter of Aldo, worked as a child with her parents and siblings to restore the land to which she retired. Nina noted that restoration is putting back what has been taken or lost from the land. She observed that restoration requires a good understanding of the land—what some have called ecological literacy, and respect for the Earth.

The first step in restoring ecological health to the land, as with administering to the health of a person, is accurate diagnosis. We are all familiar with the process. The medical staff collects family and personal history, considers symptoms, and gathers data that can be related to functionality: temperature, urine and blood chemistry, blood pressure, and so forth. Careful observation is an important part of this process. Determining ecological health is a parallel process, but it is far more complex. Unlike a human patient, the land cannot tell you how it feels. Moreover, an ecosystem does not have easily measured indicators of health like blood pressure or temperature. Thus, it is up to you to gather the history and make the appropriate observations for a correct diagnosis.

Ecologists have developed a good understanding of how ecosystems work. The processes, such as soil development and maintenance, succession, regulation of populations, and nutrient cycling are based on the interactions of literally thousands of species (most of which are invertebrates and microorganisms). It is difficult, at best, to measure these processes in any meaningful way, so measurements focus on the condition of "indicators" of ecosystem health. There are many indicators, each revealing only a tiny perspective of the whole system. This complexity, and the difficulty in assessing the condition of an ecosystem, might seem hopelessly challenging, especially for someone with minimal training or experience. Take heart from the observation that some farmers, with little formal education, have managed their farms well for decades. Ecological restoration has much in common with farming. Both require careful attention to soil and water, invasive species, and health of desired species. Having a good business plan and good management skills is important for both. This book guides you toward the observations and ecological functions you need to observe in your land, and steps for developing a sound restoration plan. Remember, if you get stuck or need help, please see appendix 4, and the website www.restoringyourland.com, on how we can help you.

Notes

1. Steven I. Apfelbaum and Alan Haney, *Restoring Ecological Health to Your Land.* (Washington DC: Island Press, 2010).
2. Aldo Leopold, *A Sand County Almanac.* 1949. (London: Oxford University Press, 1949).

INTRODUCTION

Whatever you do, or dream you can, begin it. Boldness has genius and power and magic in it.

Johann Wolfgang von Goethe

We introduce the process of ecological restoration with a brief overview of the ten-step process that has proven to be consistently successful in all kinds of ecosystems throughout North America and beyond. This approach is based on years of experience, ours and that of colleagues, with thousands of projects around the world. Although you should approach it systematically, as we have outlined it in this book, it is not a linear process. You may have to loop back through the diagnosis and analysis steps many times over several years as you gain insight or run into new problems or questions. Consequently, you may engage some of the ten steps simultaneously. This is mostly because the complexity of ecosystems inevitably leads to uncertainty in the diagnosis and, consequently, in treatments. You also will continue to learn as you implement treatments, leading to refinement of your restoration plan. The return of land to a healthy condition typically will take a few years, and there is plenty of opportunity to review and revise your restoration plan. This ten-step process may appear simple, but as you will see when we get into details, few things are certain or absolute. Ecosystem restoration requires an adaptive management approach.

The sequence of the ten steps is important, although in simple projects you can initiate several steps at the same time. Avoid skipping steps. We break most steps into a series of tasks. These, too, should be undertaken sequentially in most cases, but some may take much longer to complete, and therefore, work will sometimes overlap. Most steps build on information collected in previous steps, so a systematic approach will yield the best results. A good restoration plan is possible only with attention to the details in each task and step.

Step 1. *Inventory and Map Your Land*

In this step you will interpret the landscape and its condition. You will identify the ecological units based on edaphic (soil, hydrology, and topography) setting, land use, and vegetation. As you progress through the tasks in step 1, there will be increasing attention to detail, eventually leading to as much understanding as possible about how the ecological units have changed as a result of human activities on the land.

Step 2. *Investigate Historic Conditions*

Exploring and mapping historic conditions in and around your project can be both enjoyable and enlightening. What did the landscape look like before it was developed? What ecosystems covered the land? How did they change as a result of activities of native peoples and early, as well as more recent, settlers and farmers? Information can come from any source, including old aerial photographs, plat books, elderly neighbors, and local libraries. Some natural history books that focus on regional ecology may also be useful.

Step 3. *Interpret Landscape Changes*

Because there is no way to be certain how ecosystems have changed as a result of human activity, this step involves the development of working hypotheses about changes and causes for them based on what you find on the land and what was present historically. How did the original landscape look, and what were the ecological processes that maintained the ecosystems? For example, periodic fire and flooding may have been important. What plant and animal species were prominent? Which are now missing, and what are the effects of their absence? Try to define as best you can the kinds of ecosystems and their relationships to the landscape and processes. This is one of the steps that you will continue over the years you work with your restoration, but careful attention to the tasks in steps 1 and 2 will prepare you to begin to see what potential the land has for restoration. This is the basis for the next step.

Step 4. *Develop Goals and Objectives*

What do you hope to achieve with restoration? How will it look? What species and conditions will each management unit support? As you do this, you will need to define management units where common tasks can be applied to achieve specific objectives. Remember that goals are often more general statements, while objectives have measurable outcomes and links to technical performance. Thus, the latter especially tend to apply to specific management units within your project.

Step 5. *Develop Your Restoration Plan*

A very specific plan is important for several reasons. First, it allows you to develop a timetable for implementation that is in keeping with your available resources. Second, it acknowledges the logical seasonal and successional timing of tasks. Finally, it affords the opportunity to start small, especially where there is more uncertainty, and scale up as you learn.

Step 6. *Develop a Good Monitoring Program*

Monitoring facilitates learning as you go. Initially, you may want to capture baseline data before restoration begins, but certainly you will need to determine how units respond to treatments to be sure you are moving each ecosystem in the direction desired. If you do not carefully evaluate how the land responds to treatments, and adjust as necessary, you can waste a lot of time and money.

Step 7. *Implement the Plan*

This follows logically from the plan that you have developed through the previous six steps. Implementation will require that you have lined up necessary materials, tools, and help, and develop a phasing schedule for specific treatments. We lead you through this preparation.

Step 8. *Maintain Good Records*

Documenting changes is primarily accomplished by implementing the monitoring plan. Records allow you to periodically summarize where you are and what you have learned. Formal reports may not be necessary in most noncommercial restorations, but sharing the project's successes and failures with stakeholders should be a priority.

Step 9. *Review the Project*

We recommend that you review the project at least annually and make adjustments to reflect what you have learned and any changes in resources available for the coming year. There will be unexpected setbacks as well as successes, and adjustments are inevitable during restoration. We also suggest patience with ecosystem recovery and not making hasty, unnecessary changes.

Step 10. *Share the Restoration Process*

Working close with nature and seeing the changes that result is exciting. Certainly, stakeholders will want to look over your shoulder, if not join in the actual work, but there are many others who will profit from sharing your experience. These include school children, Scouts, and neighborhood organizations, such as garden and bird clubs. Many may provide additional hands for routine tasks.

How to Use This Book

The ten steps are a framework to organize the process of restoration. Most steps are divided into a series of specific tasks that are explained and illustrated with examples. In each step, data forms are presented for you to complete as you progress through each task. We have collected all forms in appendix 1. We also included a "crib sheet," which outlines the essential process called the "Restoration and Management Planning" data form. Data forms are provided to help you organize your investigation and planning. These forms provide an outline within which you can summarize the information you derive from the tasks in each step.

We want to emphasize one other important point that we have learned during our thirty years of managing and restoring ecosystems. When working with nature, one must have humility. Simply going through a planning process is empowering, but nature will throw you curves that no one can anticipate. Remain humble and be flexible. You likely will become discouraged at some point. Nearly all of us do. But be prepared to be awed as well. Perhaps the hardest lesson for many is learning to work *with* rather than *against* nature.

We recommend reading our previous book, *Restoring Ecological Health to Your Land*, first, although this book is designed to stand alone. Each chapter in the first book, and step in the second, builds on the preceding information to help you develop a complete and practical restoration plan. In this book, each step is detailed with specific tasks and reference to forms and tools needed to complete them.

Remember that the forms and tools provided in this book are also available online. You can download them from the www.restoringyourland.com website. The table of contents will give you the big picture, and may help you better track where you are in the process.

Staying Safe and Comfortable in the Field

Field mapping and other work outdoors should be enjoyable. Undertaking any work requires the proper equipment. Without it, productivity decreases and discomfort increases, usually leading to poorer results. Appendix 2 includes suggestions for basic equipment, safety, and protection for restoration planning. We make recommendations, but it is up to you to choose the right equipment to fit the unique aspects of your project, climate, and experience. If you are taking others, especially inexperienced people, you need to ensure that they will be safe and comfortable. Should an emergency arise, know proper procedures. We cannot overstress the importance of being properly prepared.

Let's get started!

Step 1.

Inventory and Map Your Land

If the land mechanism as a whole is good, then every part is good, whether we understand it or not.

Aldo Leopold

Ecological restoration must begin with understanding the land with an emphasis on determining its ecological health. Rarely will a tract of land have been sufficiently studied at the outset of restoration planning to provide a sufficient assessment to complete the restoration and management planning data form (see appendix 1). More typically, you will start from scratch. This step begins with an overview of how to assess the land then covers the details for evaluating the ecosystems.

Land evaluation involves three processes that can be done sequentially, or more or less simultaneously, depending on the size and complexity of the landscape:

1. Identify and map the ecological units (plant communities or cover types if the land is not too disturbed or in agricultural crops).

2. Characterize the communities as to dominant species and location in the landscape with special attention to ecotones and succession.

3. Determine the ecological health of each community or ecosystem. Note: we will refer to either "community" or "ecosystem," depending on whether the emphasis is on the species pres-

ent (community), or the ecological processes (ecosystem).

Each process demands more knowledge and interpretative skill than the previous. Some steps may exceed your knowledge or experience. If so, we offer alternatives but encourage you to keep learning and seek assistance when needed.

The first process requires only good observation to delineate the different ecological units present. We use "unit" here because disturbance often disguises the variation in edaphic conditions such as soil, hydrology, and topography that would have led to variation in vegetation patterns. Where vegetation patterns are present, they reveal ecological units. For example, you may see a weedy hillside grading into a wetland at the bottom of the slope. These are different ecological units that you would map. You need not worry about the details of species and variation in soils at this point, but simply look for differences in vegetation, topography, or hydrology that are apparent. Smaller fields, lawns, or vacant lots will generally be mapped as a unit at this stage, especially if you see no natural breaks in topography or hydrology.

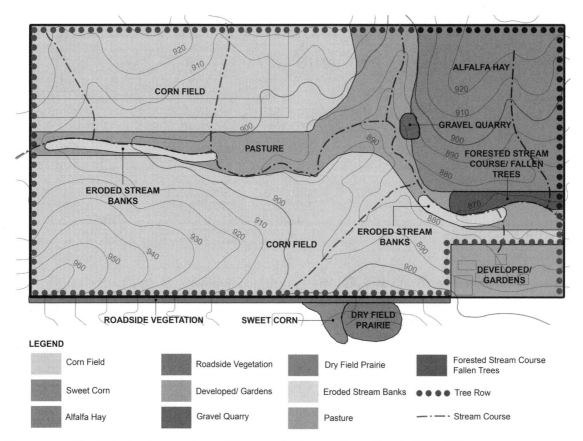

FIGURE 1.1. Ecological units on a simple Wisconsin landscape, Stone Prairie Farm, WI

The aim of the second process is to define and characterize the ecosystems that occupy each unit you have identified. The dominant plants are most often used, but variation in soil and hydrology may also become essential clues, especially if the vegetation has been largely altered by agriculture or development (fig. 1.1). This will require that you investigate deeper, perhaps examining soil characteristics, although superficially at this stage. During this process you refine your initial interpretation of the landscape.

The third process is the most demanding. It involves "detective work." In this process, the aim is to understand the reasons for the patterns you have mapped. What are the differences between the units? Are the differences a result of variation in soil, hydrology, or human disturbances? Perhaps they represent different successional stages following some prior disturbance. How was the site disturbed? How do units grade one into the other along slopes, shifts in soil characteristics, or hydrology and drainage patterns (fig. 1.2). Answers likely will involve variation in edaphic conditions, but also past and ongoing disturbances, what are called stressors (fig. 1.3). Understanding how past and present stressors shape ecosystems is fundamental to the development of a good restoration plan.

On a very simple piece of property, you may be able to conduct all three processes at one time.

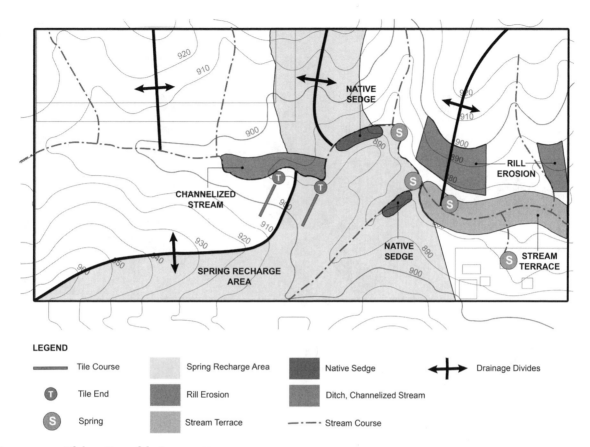

LEGEND

Tile Course	Spring Recharge Area	Native Sedge	Drainage Divides
(T) Tile End	Rill Erosion	Ditch, Channelized Stream	
(S) Spring	Stream Terrace	Stream Course	

FIGURE 1.2. Elaboration of drainage patterns

However, this is not possible on complex properties. A simple landscape may have only one edaphic setting that now is cultivated, perhaps a corn or wheat field, or a vacant lot. In most cases, however, even a forty-acre, nearly flat field will have variation in soils and hydrology worth noting. A complex landscape will contain many ecological units, perhaps with remnants of historic vegetation in various conditions of ecological health comingled with cultivated land, and often other units that have been altered in various ways. Commonly, hydrology also will have been altered by channelization or with agricultural tiles.

Regardless of whether simple or complex, the three processes are equally important. The easiest way to capture the information is by mapping what you see on the landscape at the appropriate scale over a topographic map or aerial photograph, a process we explain next. The maps become the basis for subsequent restoration planning, so care in developing them is important.

Mapping

The information required to develop good restoration plans must be georeferenced, that is, the information is site specific. Each item of information relates to one or more specific locations on the land. Maps are used to georeference the

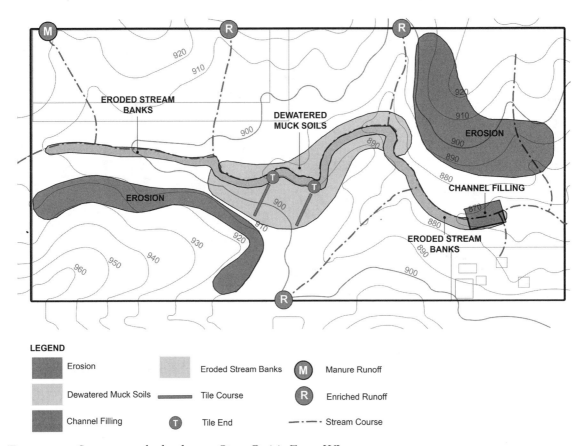

FIGURE 1.3. Stressors on the landscape, Stone Prairie Farm, WI

information and facilitate collection and compilation of information, as well as guide restoration treatments and monitoring to be described later. All tasks associated with land evaluation involve mapping what you find in the field. Mapping will require that you have some basic tools and equipment, such as clipboards, compass, notebook. (See appendix 2 for information on equipment).

It is possible to create a map of ecological units, also called an "existing condition" map, from good aerial photographs and other public domain information such as topographic and soil maps. However, we emphasize that there are several reasons why evaluations done remotely

should be considered only preliminary. Aerial photographs are often several years old and may no longer be accurate or applicable. Even with a current photograph, unless you have considerable experience in evaluating and interpreting this kind of data, it is unlikely that you can create maps remotely that will be accurate enough for the next steps in the process. Even more important, the most successful restoration efforts result from an intimate familiarity with the landscape, where you begin to understand the nuances of the land. We use the term *land* here and elsewhere in the same sense that Leopold did—the soil, water, flora, and fauna that inhabit it. This level of intimacy occurs only by being on the land. You cannot gain it from photographs or through the windows of your pickup. Required insight will come only from close observation, feeling the soils, taking measurements, learning the plants and animals, and awareness of how the ecosystems change season to season and year to year.

It is important to systematically record what you see and learn about the land. We urge that you get in the habit of carrying a field notebook supplemented with the standard field notes data form (data form 1.2, appendix 1). This form provides a catalog of features (data form 1.3, appendix 1) you should consider coding and mapping as you progress through the field assessment.

The field notes data form also leads you through a sequence of questions (box 1.1) that you should apply to each ecological unit you identify and map during the field inventory. As you map an ecological unit, either go through the field notes data form at that time, or return later to do so. When you are finished with this initial evaluation, you should have a field notes data form page completed for each ecosystem or land unit defined in the hand sketch.

The balance of this step leads you through these processes task by task. The more carefully you complete these tasks, the more likely you are to develop a successful restoration plan.

Field Assessment

If you are not well acquainted with the land to be restored, we suggest you begin by reviewing a plat map of the property usually obtainable from the Natural Resources Conservation Service (NRCS) office or from the county clerk of deeds. Familiarize yourself with property boundaries and walk the land to get a feel for topography, streams and wetlands, disturbances, and general vegetation distribution. This is facilitated by comparing what you see in the field with a soils map, topographic map, or aerial photograph, preferably all three. You will need these layers of information for your basemap anyway, so better to have them from the outset. Topographic maps are readily available from the internet through TerraServer.com or topomaps.usgs.gov/. At Terra-Server you can download either aerial photographs or topographic maps covering any property in United States. You need to reconcile property boundaries with the aerial photograph or topographic map, however. Your NRCS office probably can provide a soils map showing property boundaries. Alternatively, for many locations in the United States, soil maps have been digitized; go online and enter "soil map" and your county and state to download a soil map covering your property, but you will then need to overlay property boundaries. Many counties also have digitized plat maps that can be downloaded.

High quality aerial photographs usually can be obtained from the NRCS office. Get the most recent flight, often no more than a few years old. Some offices may have photographs dating as far back as the 1930s, when aerial photography was just being started and these older maps will also be useful when you investigate the history of the property (step 2). Ideally, it would be good to have a historic photograph from each decade. Either have the photographs scanned at very high resolution, or reproduced using a photographic service. If you cannot find the photographs locally, they can be ordered online for most areas in the United States. Check National Aeronautics and Space Administration (NASA) (http://earthobservatory.nasa.gov), TerraServer.com, or GoogleEarth (http://earth.google.com/).

Box 1.1. Questions to Help Become Familiar with Land Assessment

Vegetation and Plant Communities

1. Are the different types of plant communities easily definable, and can you map them clearly and distinctly?

2. Where are there difficulties in defining plant communities, what and where are the confusing features?

3. Is the confusion resulting in a location where one type of plant community may be invading another or where a recent disturbance obscures site condition?

4. What would happen if growth of the plant community and disturbances continued unbothered in, say, ten to twenty-five years? Would your mapping of the types and boundaries of each plant community type change?

Physical and Chemical Features: Bedrock, Soils, Hydrology, Nutrient Enrichment

1. Are there distinct transitions in any underlying physical setting that are reflected by the plant communities, including crop growth and success?

2. What changes in the site's physical or chemical factors are believed to have occurred or continue to operate on the land?

3. Are there locations on the land where a reduced availability of light by dense invasive vegetation growth is contributing to the decline of soil-stabilizing vegetation, and soil erosion is occurring or possible?

4. Are there locations where groundwater emerges to the surface as springs, seeps, marsh, or wet ground? Based on land forms are there areas where the surface infiltration is likely to be occurring to support the springs or seeps?

5. Are there locations where soils move, slump, erode, or are unstable because of frost heaving, shrink-swell?

6. Are there locations where runoff from neighboring lands enters your land, or is your land receiving soil eroded from a neighbor's land? Farm or ranch land, or urban or industrial setting?

7. Have drainage features been modified by straightening, deepening, damaging, draining, drawing, or diversifying?

Emotional, Spiritual, Inspirational, Experiential

1. Are there locations that excite you because of existing beauty, or features that cause wonder or awe?

2. Are there locations you associate with some historic experience—good or bad—or stories told to you by others when you have some sense of historic conditions and changes?

3. Are there locations that disgust you emotionally, when you get concerned about what is happening to the land, or when you feel helpless to address existing concerns?

Historic Sense and Story

1. Are there locations where you are reminded of something you or others observed—some previous wildlife encounter, a family event?

2. Are there locally or regionally associated historic records pertaining to your land?

3. Do local arrowhead hunters, fisherman, mushroom hunters, or photographers know about your land?

Using either an aerial photograph, topographic map, or soil map on which you can identify property boundaries, return to the field to sketch ecological units as previously described. Try to capture the natural variation in the land based on topography and hydrology, reflected by vegetation, if relatively undisturbed. If the land has been farmed, these units will commonly correspond to fields, where variation in edaphic conditions was incorporated into farming practices. In larger fields, however, you may see variations that should be mapped. If areas are relatively undisturbed, map according to dominant species or vegetation type. This is often called cover-type mapping, each cover type corresponding to an ecological unit. On small properties, this mapping can be done in an hour or two, but on large properties, it might take several days.

Task 1. Create a Basemap

Once you have a good overview of the land, you are ready to prepare your basemap. Digital basemaps are much more convenient and will save you a tremendous amount of time over the course of your restoration project, especially for larger projects. Even if you do not have access to a computer, you may be able to get a computer-savvy friend who does, or someone you can hire to create a basemap for you. Or, if you wish, contact our website to ask for our assistance. Alternatively, go to the NRCS office and ask for help in creating your basemap. As a last resort, see if a teacher at the local high school or technical school would be interested in using the preparation of your basemap as a class exercise, perhaps even staying involved in the restoration process. We will describe both the hardcopy and digital

processes, but we urge that you seek assistance as necessary to produce a digitized basemap. In any case, the essential layers of information you will eventually need on your basemap include the following:

- property boundaries
- soil types and distribution
- topography including streams and water bodies
- landforms such as rock outcrops and wetlands not otherwise shown by topography
- roads and utility lines

Any time you can inexpensively increase the precision of your efforts, it will pay off in the short as well as the long term, regardless of the size of your project. Especially on larger projects, errors in mapping can become costly.

Preparing a hardcopy basemap. Because maps are often at different scales, you likely will have to reduce or enlarge maps to match scales as accurately as possible. We recommend you use the United States Geological Survey (USGS) topographic map as the ground layer to build your basemap. In the rare case where you have another topographic map at a finer scale, as long as it covers your project site and some context of land beyond your project boundary, then use it.

Most USGS maps are at one of two scales, a 7.5-minute or 1:24,000 scale where one inch on the map is equal to 2,000 feet on the land, and a 15-minute or 1:250,000 scale where one inch on the map is equal to 20,833 feet on the ground. Because these scales will result in small properties being just a few inches square, or less, we recommend that you enlarge the images to fill much of an 8.5- by 11-inch page, and recalculate the scale so that when you take measurements in the field, you can relate them to your map. Use a photocopy machine to blow up the image.

Initially, the aim is to create a basemap on a piece of paper that is large enough for you to record boundaries of what you find when you are in the field. Scale is less important, as long as it is convenient for your use, and you do not have to write so small that the mapping becomes illegible. If so, increase the resolution of your basemap either using a photocopy machine or, if digitized, change the scale with the computer.

Basemaps of two scales are desirable, one showing a broader landscape to provide a context for your property, and the other that focuses specifically on your property.

Smaller map. This map should focus on your property or project site and at least some of the adjacent properties. For the authors' farms, these maps included a quarter-mile border beyond the project sites.

Larger map. This map should provide the context of your project site. For the authors' farms, we included a border of two miles beyond the project sites. The landscape context you choose is determined by the maps you have available and potential impacts of external stressors. This map can have less resolution, often using the 15-minute topographic map.

When Steve began restoration of his farm, he wanted both the topographic and soils information on the same basemap. Both the published USGS topographic map and NRCS soils map were at the same 1:24,000 scale. His farm was eighty acres, but covered little more than two inches in one direction and one inch in the other direction on these maps. He enlarged both maps using a photocopy machine so the farm, with a quarter-mile border, covered an 11- by 17-inch page. He then recalculated the scale which turned out to be about ten times the original, roughly one inch to two hundred feet on the ground. Although that is an ideal scale, if your project is much larger, you may need to work with less resolution.

Using tracing paper, soil types from the enlarged NRCS soils map were transferred to create an overlay of the topographic map. This was then traced onto a clear Mylar plastic film material (available from office supply stores) using a Sharpie pen. Once this was done, the Mylar film with the soils boundaries was taped over the topographic map, making sure all property corners aligned, and a photocopy machine was used to make a combined map. Duplicate copies were made as needed for the field, keeping the original in the office.

This same technique was used with a topographic map to compile topographic lines, streams, buildings, roads and others features onto the basemaps. This was combined with the soil-type boundaries over a recent aerial photograph that was enlarged to the same scale. This resulted in a refined basemap with all three layers on a single map. Of course all this could have been done much easier with geographic information systems (GIS) technology, which was not available when Steve began his project.

Preparing a digital basemap. With computer-assisted mapping tools, handheld global positioning satellite (GPS) units, and the growing availability of GIS technology for nonexpert users, nearly anyone can make basemaps, develop your restoration plans, and keep records of management and restoration activities digitally. This will facilitate planning and management decisions as well as monitoring (step 6). The USGS website allows you to download the latest topographic maps (http://store.usgs.gov). Alternatively, you can obtain topographic maps from sources such as DeLorme TopoUSA (http://www.DeLorme .com), which allow you to customize and print maps of your property.

The NRCS soil survey website (http://www .nrcs.usda.gov/) allows you to identify your approximate property or project site boundaries online and create a map that you can print or download. This will show the boundaries you define, the soil-type boundaries on the property, and will give you information on soil types and their characteristics. Be aware that these soil maps are rough and may need to be refined, as discussed later.

If you are comfortable using computers, there are many internet sites where you can access computer-aided mapping tools that will be useful. If you are not proficient with computers, however, get assistance until you gain some comfort with digital processing. Some of the natural resource–focused sites include:

1. http://www.tatukgis.com
2. http://datagateway.nrcs.usda.gov
3. http://forestpal.com
4. http://maps.dnr.state.mn.us/forestry /photos/
5. http://www.isgs.uiuc.edu/nsdihome /ISGSindex.html
6. http://megi.state.mi.us/mgdl/
7. http://igsb.uiowa.edu/nrgislibx/

There also is a wealth of GIS technology that allows you to download digital layers of information, such as topographic and soil distribution. There is free GIS map-viewer software with which you can customize your own basemaps. Using this same software, you can record all future land management and restoration activities, and create data layers associated with each activity in your project. You can also save monitoring data and digital photographs to create permanent records or databases easily accessed in the future by you or others. This software permits powerful manipulation of information, such as comes from interpretation of multiple variables achieved by overlaying layers of information. GIS also allows users to do analysis. For example, if you have geo-referenced (using GPS) locations of each population of a rare terrestrial plant, you can use the GIS software to predict other possible locations where additional populations may be found, based on the patterns and characteristics of locations where it first was found. GIS also supports change analysis. For example, if you mapped your project site today, you could reproduce the mapping process in the future and then have the GIS software map changes in boundaries between the two mapping periods. The GIS will provide revised acreages and related information to support restoration planning. On complex projects, consider contracting for GIS services to gain this level of sophistication.

Avoid getting bogged down with technology that exceeds your experience. That is one reason we offer the online consultation mentioned in the preface. For example, if you wish to do a digital basemap, but have trouble creating your own, please let us help you.

Whether you do a digital basemap or a hard-copy, when you go into the field, you will need extra hard copies for reference and marking up. This is easily done by printing copies off a digital basemap or making photocopies of your original hardcopy. With a digital map and minimal experience, you also can enlarge or reduce the map to fit your needs. Otherwise, you will need to use a photocopier to change the size.

Task 2. Characterize Land Use or Cover Types

In this task you will direct your attention to what is growing on each unit or how the land is being used. You can think of this exercise either as mapping what is there or mapping how the land is being used. Categorize units into broad categories, such as annual row crops, orchards, pastures, forests, residential building zones, wetlands, lakes, roads, and other areas such as waste disposal.

Extend the land-use map to neighboring properties within the limits created in your larger basemap. Using extra copies of the basemap (keep the master copy protected), write and draw boundaries directly on the map. If you prefer, you can tape a piece of transparent Mylar to the surface of the map and use Sharpie pens to map over the basemap. If you use Mylar, draw registration marks on the overlay that align directly over at least three permanent features on the underlying basemap so that you can quickly realign the overlay once you remove it.

As you map, you will be recording several kinds of information. Data forms and maps become the record-keeping tools for each trip into the field, but you should also carry a camera to photograph cover types or other features. GPS units are especially handy to georeference features or photographs. They have become relatively inexpensive and easy to use, and can save you many hours of field time while increasing accuracy. Additional forms for recording field notes include the photograph record data form (data form 1.4, appendix 1). GPS record data form (data form 1.5, appendix 1), sample collection data form for soils, plants, etc. (data form1.6, appendix 1), and species Listing data form (data form 1.7, appendix 1).

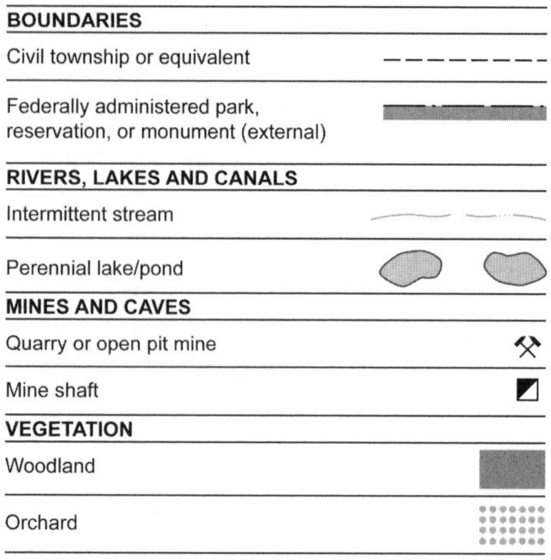

BOUNDARIES	
Civil township or equivalent	– – – – – – –
Federally administered park, reservation, or monument (external)	▬▬▬▬
RIVERS, LAKES AND CANALS	
Intermittent stream	– – – –
Perennial lake/pond	⬭ ⬭
MINES AND CAVES	
Quarry or open pit mine	⚒
Mine shaft	◨
VEGETATION	
Woodland	▮
Orchard	⣿

FIGURE 1.4. USGS mapping symbols and codes

It will be useful to standardize codes or information you place on your map. We recommend the standard symbols and codes used by USGS in their mapping (see examples in fig. 1.4, or access the website for USGS at www.usgs.gov). Doing so will save you time by not having to record what your symbols mean.

The USGS codes may not cover all types of land uses you want to identify if you wish to map with greater specificity. For example, you may choose to differentiate an irrigation pond or livestock watering pond from a small natural lake. Be sure to define them well on your maps. Keep it simple and make sure that whatever you use on the map, you use to map similar features throughout your project. Keeping a field notebook with these details will be important in addition to information you record directly on the field maps or on data forms.

Standard land-use mapping is routinely done in the United States by USGS, The Nature Conservancy (TNC), and state, county, and municipal agencies.

Categorize the land use or cover types in each ecological unit you previously identified in the field. On simple projects this may require only an hour or two; on complex landscapes, it can take weeks of field time.

Task 3. Refine Ecological Units

In this task we want to characterize the units into refined categories. Where fallow fields or pastures were previously mapped, now differentiate the vegetation types present in each area. For example, separate cultivated crops into the type of crop, and forests into cover types, based on dominant species. Differentiate as completely as possible. In smaller projects, you may be able to map individual plant communities, perhaps no larger than a few square meters. In larger projects, you will be able to map only higher levels of plant associations. Some of the categories and symbols used in task 2 will be the same (e.g., pine plantation, pond, etc.).

Vegetation classification. In mapping cover, think in terms of a hierarchy. For example, the region might coarsely be classified as tallgrass prairie or oak-hickory forest. Within it will be many small vegetation types, such as floodplain forests, different types of wetlands (box 1.2), and all types of disturbed communities from recently cultivated fields to pasture. Within each of these, there will be specific plant communities that reflect the variation in edaphic conditions. If the natural vegetation is still largely intact, it will be useful to know what regional and local associations have been mapped and classified, and follow that system as much as possible. These may not resemble associations that have been previously described for your area, but knowing what was there previously is still very important. This is discussed further in step 3.

Original vegetation is the vegetation present prior to the alterations of the land. In regions such as Europe and the Middle East, where land and vegetation have been altered for centuries, about all one can do is project the likely plant association that occupied each ecological setting. Prior to the sixteenth or seventeenth centuries in the United States and much of Canada, Native American and First Nation people influenced fire regimes and game populations, and in some instances, these effects were quite profound. Thus, the common use of the terms *original vegetation* or *presettlement vegetation* is somewhat misleading. Original vegetation maps are typically developed by interpreting General Land Survey records completed by the government surveyors in the late 1700s. It was this survey that established the grid of one-square-mile sections, which now cover most of the United States, shown on the USGS topographic maps. Every landowner knows this land subdivision, or at least sees property-tax bills that reference the land subdivisions by townships and ranges.

Existing vegetation refers to vegetation present on your land at the time of your survey. In most parts of the lower forty-eight states of the United States, the original vegetation was altered long before aerial photography was begun. Nevertheless, the use of older aerial photographs can be helpful to map communities or land cover types. The resulting maps are often called *historic vegetation cover maps*, but they are usually not original vegetation. In some parts of the United States that were settled very late, it may be possible to find historic aerial photographs or maps that show original vegetation, but that is unusual.

On land cover maps, be aware of the scale of interpretation during mapping. Standard classification systems and dichotomous keys are available for many areas of the United States, at least at

BOX 1.2. EXAMPLE OF A DICHOTOMOUS KEY FOR LAND COVER

1.

    a. Land perennially dry or moist but never inundated with standing water, always elevated or level and well drained. Plants erect and supporting themselves or growing intertwined with other upright plants . . . . .2.

    b. Land inundated with water at least seasonally; soil saturated at least part of the year; plants are floating, submerged, or shallowly rooted in substrates, often unable to support themselves in an upright growth form, without the buoyancy of water . . . . .7.

2.

    a. Woody vegetation dominant, although it may be restricted to narrow bands of tree cover along waterways . . . . .3, forest and savanna.

    b. Nonwoody vegetation such as grasses, sedges and herbaceous plants dominant. These may be agricultural lands, pastures, or wild lands . . . . .4.

3.

    a. Tree cover > 50% , comprising several size classes of trees; several tree species are present. A dispersed shrub and sapling layer is more or less uniform beneath the tree cover. . . . . .5.

    b. Tree cover < 50%, and over half the area is dominated by herbaceous plant cover of grasses or forbs. Where present, trees are small, often growing in tightly clustered thickets, sometimes beneath larger mature trees . . . . .6.

4.

    a. Agricultural land; herbaceous plants in noticeable rows, or harvested for forage, or pastured . . . . .annual or perennial cultivated or harvested grasslands.

    b. Grassland or herbaceous oldfields not recently planted or harvested, but may be seasonally pastured by grazing livestock . . . . .perennial nonharvested, or pastured grasslands.

5.

    a. Topsoil < 6 inches, brown (seldom black); fallen trees in various states of decomposition. Often, tree-sapling layer is of mixed species, including shade-tolerant oaks, maples, and ash of various size classes . . . . .deciduous forest.

    b. Topsoil typically 6–12 inches, very dark or black; scattered large oaks with dense understory of introduced shrubs, such as European buckthorn and Tartarian honeysuckle, and invasive herbaceous plants such garlic mustard. Understory trees of species different from overstory, such as Norway or sugar maple . . . . .degraded oak savanna.

6.

    a. Topsoil mixed with subsoil and light brown in color; ground layer dominated by introduced forage grasses such as European brome or tall fescue, and such plants as Canada goldenrod . . . . .former agricultural land (oldfield).

    b. Topsoil typically 6–12 inches deep, primarily black in color. Scattered large oaks with dense understory of introduced shrubs, such as European buckthorn and Tartarian honeysuckle, and invasive herbaceous plants such garlic mustard. Understory tree species different from overstory . . . . .degraded oak savanna.

Box 1.2. Continued

7.

    a. Water primarily open year round, with a sandy or gravel beach downwind of the prevailing wind direction . . . . .lake.

    b. No open water for portions of the year, and area more or less completely overgrown with vegetation; beach lacking . . . . .8.

8.

    a. Depression is covered year round by > 50 % herbaceous vegetation rooted in the underlying substrates. Vegetation may die back in winter and reemerge, standing erect by midsummer . . . . .wet prairie or marsh.

    b. Water body seasonally open with herbaceous vegetation emerging only during dry years or dry seasons . . . . .aquatic.

---

the regional scale. They may help in understanding the mapped categories you might use in preparing maps for your property.

Dichotomous classification key. Taxonomic keys have been developed to aid identification of organisms. Similar keys are available to help classify land cover or vegetation types. Figure 1.5 illustrates such a key, which we have developed to introduce the concept. For each pair of criteria, choose the one that best fits the vegetation and progress to the next set of couplets. Using a key like this is not necessary if you feel comfortable recognizing cover types. Study the key, however, to get a sense of the level of specificity desired.

Classification of land cover has been developed by many state agencies, often through a natural heritage division. You can also find these at NatureServe (www.natureserve.org), through The Nature Conservancy, and other conservation organizations. The NatureServe site is comprehensive, with excellent information, but it is not very user friendly, so we walk you through an example for using it here. For practice, we will lo-

cate the association for the deciduous forests on Alan's central Wisconsin farm. The forest is dominated by red oak, white oak, black oak, and red maple with bigtooth aspen and a small amount of sugar maple.

1. Go to the NatureServe website at www .natureserve.org.

2. Click on "NatureServe Explorer."

3. Below the "Species Quick Search" box (which should be left blank), click on "Ecological Communities & Systems."

4. When the new page comes up, click "Associations" under the "Search by Name" heading.

5. Scroll to the bottom and click on "Location" to narrow the search.

6. When the list of states comes up, click on Wisconsin, then on "Search Now."

7. A list of Wisconsin associations will come up. Scroll through them to "Deciduous forests," and under that the association with red oak (*Quercus rubra*) and white oak (*Q. alba*) as the two dominants.

8. Click on this association and it will give you

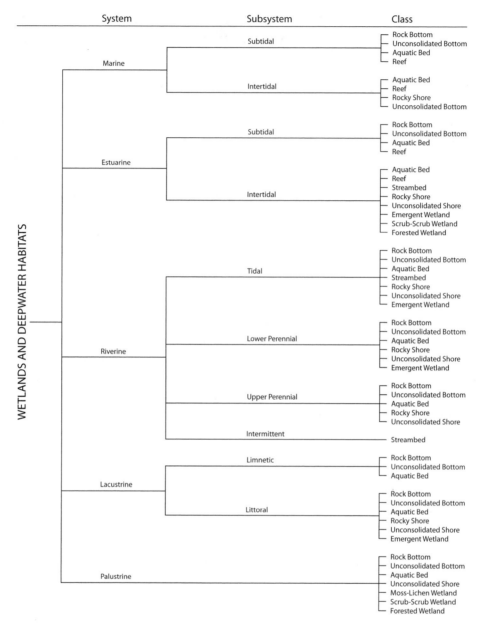

FIGURE 1.5. Example of a dichotomous classification key

a good description, including soils, landform, and associated species. It is an excellent fit.

Try working through with an association in your own project if you have one that is dominated by native species. The associations de-

scribed in NatureServe are what would have been present before they were removed or destroyed on the site.

You should now create a current land cover map for your land. Take a fresh look at the land and vegetation. Prior to actually mapping, create your classification. You can either adopt categories from a land cover map from your area, with modifications as necessary, or develop a classification based on the categories of vegetation cover you find as you walk over your land. Pay some attention to landform, as there may be important differences depending on drainage, slopes, and soil, even where agricultural disturbance has disguised edaphic variation. Hold off on actually mapping until you have a clear idea of the categories of vegetation cover you will map. This also would be a good opportunity to photograph each land-use or cover type you will recognize. The photographs will provide some baseline documentation of the prerestoration landscape.

Often it is easiest to develop your classification in response to what you find. This requires, however, that you know enough to consistently identify different cover types, and resolve those differences to a useful level. In other words, do not try to split hairs, but also avoid grouping into categories that are too large. Whether you follow their lead or not, knowing how others differentiated local cover types can help. Also, standardization of the nomenclature and map symbols can save you time in the long run. The amount of time you spend walking your land, observing, and thinking about categories will depend on your experience and familiarity with local vegetation. Do not cut this process short, however, even if several field days are necessary to get started.

In addition to a basemap with topography and soil overlays, we suggest you carry descriptions of the cover types you have decided to recognize. With experience, you may not need them, but at the beginning this can increase your consistency in mapping. If you have not previously done so, record descriptions of each cover type you map in sufficient detail so that you can consistently recognize them. Map each unit on a copy of your

basemaps. Most of the same equipment you carried during the field work for task 2 will be needed.

Assessing Conditions on the Land

Now that you have mapped the landscape components, we are ready to become more analytical. Through the following tasks, you will develop a deeper understanding of the ecological conditions on your land and how the stressors led to them. This work, which will continue through the life of the project and beyond, begins what for most landowners is a life-long intimate relationship. As Thoreau implied, the complexity of nature is so great that we can never in a lifetime reach into its most inner secrets, and it is that mystique that makes nature study so consuming. This process requires more time and more study, and the more you learn, the better your restoration is likely to be, but because complete understanding is beyond even the most dedicated ecologists, you should not be too critical of your own shortcomings. Delve as deeply as your energy and time permit, and move on, but return as frequently as possible to puzzle the patterns and intricacies of nature.

Task 4. Map Current Conditions of Ecological Units

Unlike the previous mapping, you are not mapping what is growing on the land, although that is likely to be an important clue. Instead, you are mapping the causes or reasons for the patterns in what is growing or occurring on the land (see fig. 1.6). This task may not result in a different map. Instead, you may simply chose to tape a clean transparent Mylar over the map from task 2 or

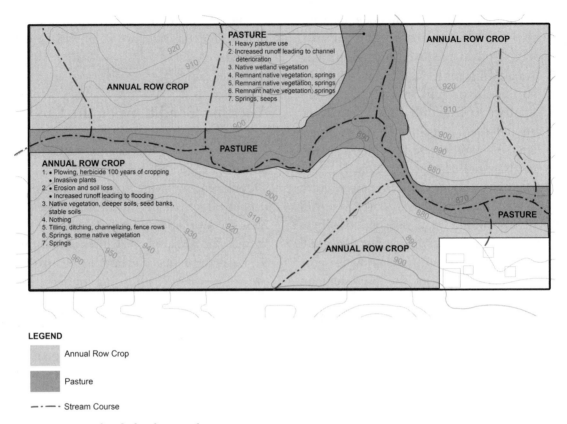

FIGURE 1.6. Example of a land cover change reasons map

task 3 for recording notes and your answers to some of the questions found in box 1.3. See data form1.8, appendix 1 for a form in which to answer the questions pertaining to ecological change reasons.

This task may be as simple as relating back to the land uses that you mapped during task 2 with cross-referencing to topography and soil maps if the cover is explained solely by edaphic conditions. Usually, however, there will have been various kinds of disturbances, often with complete elimination of any trace of natural vegetation. This is where you must try to understand what has happened to the land.

Agricultural crops are often the most problematic. Modern agriculture even tends to over-

whelm subtle variations in soil and hydrology. With equipment, chemicals, and cultural techniques, farmers have managed to convert wide ranges of landforms and soils to agricultural production. The less disturbed the vegetation, the easier it is to recognize relationships with landform and soils. Take, for example, a cornfield—if, as is usually the case, the corn will be grown on two or more soil types. Using the soil survey map, can you see subtle differences in the corn on different soils? Ask the farmer. He may be more aware because of yield differences, but he may also be compensating by using different fertilizer and herbicides that mask differences. If corn is growing on hydric soils, those formed in wetlands, it is very likely the soils have been tiled to

---

BOX 1.3. UNDERSTANDING CHANGES IN THE LAND

1. What conditions gave rise to the vegetation patterns?

2. What changes or stressors have contributed to existing conditions?

3. What is missing from what would have been here before settlement?

4. What is present that would not be expected on this landscape?

5. What evidence is present that indicates how the land has been modified or altered by present-day, recent-past, or earlier humans? Insight into people that lived on the land prior to records may be important for understanding the history of the land. Watch, for example, for places where hand-laid rock weirs may have been placed hundreds to thousands of years ago into drainageways, or where some usual vegetation patterns or features (rock piles, old building foundations, or geometric patterns) are present.

6. What is different on my property compared to neighboring land?

7. What features have I been told about or heard about on my site that others consider to be unusual, unique, or for which there is uncertainty?

---

remove excess water. Often there will be nearby drainage ditches. In mapping fields that have been tiled, you will need to watch for subtle clues such as we later describe in task 5.

Wetlands may be disguised not only by dewatering with tiles and ditches, but also with dredging and filling, widely practiced until the Clean Water Act was passed by Congress in 1972. Consequently, investigate low areas carefully. Over two dozen kinds of wetlands are recognized around the world. Those with standing water and emergent vegetation are obvious, but some, such as sedge meadows or ephemeral wetlands, may not be, especially if they now are being used to grow crops. In the United States, the National Wetland Inventory has mapped most significant wetlands and many states have expanded this inventory. Generally you can get these records online through your state department of natural resources website. If a wetland is indicated on your property, cross-reference it with the appropriate soil survey map. Soils under wetlands are hydric,

meaning that they have insufficient oxygen at least a portion of the year, resulting in reduction of iron compounds. Reduced iron compounds are blue to gray whereas oxidized iron compounds are bright yellow to reddish. Hydric soils are nearly always very high in organic matter, as well. This gives them a dark color. To confirm a hydric soil, use a shovel or soil probe to reveal soil down at least a foot or more. If at least the surface soil is dark, and the soil beneath the surface a foot or so shows areas of blue to gray mottles, it is very likely the area once supported a wetland of some type. If natural vegetation is present, at least some of the species will be characteristic of wetlands. In highly disturbed wetlands, it may take awhile to probe the suspected margins of the former wetland sufficient to map the extent of the wetland, but pay attention to topography. The wetland margin will likely follow the elevation contour, so once you have located a margin, you have a good clue for how to map the entire wetland, and can proceed to spot-check your projection.

In western rangeland, dense growths of young cedar or juniper trees, or vast areas of introduced cheat grass (*Bromus tectorum*) are symptomatic of both problems and significant changes on the land. The presence of cheat grass, as well as other invasives, affects the ecology. For example, native grasses and forbs will not persist under dense juniper. Cheat grass's aggressive spread and early spring germination and growth can out-compete native grasses that require warmer soil temperatures for germination and growth. The absence of native grass can result from a number of reasons. This includes competition, but often more important, cheat grass is dominant where soil compaction, erosion, and sedimentation have occurred, commonly as a result of heavy grazing. Thus, the presence of cheat grass is not only a direct deterrent to native vegetation, but may also indicate an underlying condition stemming from an overgrazing stressor that may or may not still be present. Try to consider both direct and indirect ecological relationships in what you map.

Identifying the underlying relationships that give rise to land-use and cover types needs to occur progressively, over time, as you learn and become more familiar with the history of your land, and its use. Unlike the previous steps, this is a long-term process. You will initially learn some reasons at this time, but others will elude you and only be understood later. Perhaps some patterns cannot be explained, but you may get an inkling of reasons from neighbors or as you do restoration on your land. There will be subsequent tasks that also will help elucidate some reasons. In locations where you have no clue as to why a particular vegetation occurs there, make a notation and leave these blank. Return frequently to the field to explore your hunches. Watch for clues such as old plow lines, former fence-rows, old roadbeds, channelized or ditched streams and drainageways, or erosion features.

Surveying Soils

Soils are a fingerprint of the past and a key to restoration. The characteristics of the soils reflect the geologic substrates, topography (including drainage), organisms (especially vegetation), climate, and the time the soils have had to develop. Past disturbances will also be reflected and can be seen decades, if not centuries, after the fact. If you have had little or no background in soil science or terminology, reading soil profiles can be challenging. Check with your county extension office to see if they or the county conservationist (through the NRCS office) can connect you to a local soil specialist who might assist you. Lacking that, there are really only two options. For most projects you can simply use the NRCS soil map that covers your area with the inaccuracies it will contain. Alternatively, you can pick up a book on soils and begin learning what you need to make this work worthwhile, seeking professional support where you can find it. One of the best references for a beginner is *Soil Science Simplified*.[1] Although more advanced, NRCS publishes a soil survey manual, available online at http://soils.usda.gov/technical/manual/.[2] Use an internet search for *soil morphology* or *NRCS soils* to get access to a wealth of related information, including definitions, pictures, and descriptions of soils and soil maps throughout the United States. Canada also has excellent resources for soils. Your county probably has a soil survey with maps and complete descriptions for each soil type. Check with the extension office to find out where you can get a copy.

Soil survey maps have minimum unit sizes, meaning that smaller parcels commonly were ignored or lumped with adjacent soil types. For example, if there is a small depression in a well-drained upland field, the small area is likely to have a different soil, perhaps a former wetland that has been ignored in mapping the surrounding lands. Most mapping was done remotely using aerial photographs with minimal ground-truthing, that is, confirmation on the ground of what was remotely interpreted from the photograph. The surveys commonly missed small or subtle landform features such as minor drainageways and depressions. Even if you bypass this task and rely on the NRCS soil map, read through the following, and at least modify the soil map by noting small features where soil is likely different but lumped in the NRCS map. These sometimes contain rare species, or offer unique opportunities in restoration. Also, either online or with the country soil survey booklet, read the description of soils in your project. Complex projects usually have many landforms, soil types, and vegetation communities, and careful field investigation can take weeks (fig. 1.7).

Soil surveys first were completed in the 1930s,

and mapping continued through the 1970s in the United States. In many locations, surveys were revised more recently. While these are generally accurate, they are not without errors and usually miss some features that may be important in restoration.

Task 5. Review Soil Type Distributions and Assess Seedbanks

This task can be divided into two phases. First you should try to understand where inconsistencies exist; the basic soil characteristics and descriptions; where soil conditions (e.g., compaction, erosion) have changed; the locations and nature of disposal areas, such as farm dumps; and nutrient enrichment or toxic areas and the nature of the chemistry.

Modify a copy of your basemap to reflect what

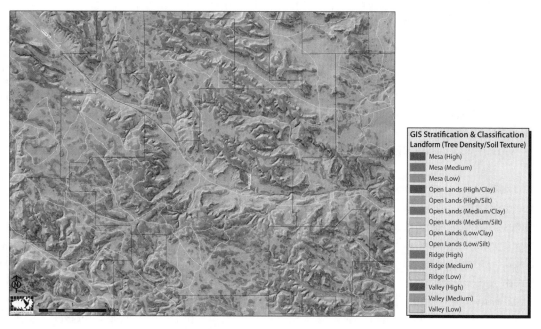

FIGURE 1.7. Example of a digital elevation model used to refine a basic soil map

you learn by sketching in changes to soil type distributions and notations or related things you observe.

Soils nearly always accumulate seeds and other kinds of plant propagules. *Seedbanks* are viable, living propagules in the soil. Some seeds can persist for hundreds of years, waiting for the proper conditions to stimulate their germination. Your goals for this phase of the task is to determine the presence and conditions of seedbanks and the comparative abundance of nonnative and native propagules. Both phases of this task involve examination of the soils following the process outlined below. Use the soil data form (data form 1.9, appendix 1) to record your observations. This task should be undertaken only when the soil is moist, such as in the spring or a few days after a soaking rain. Subtleties are difficult to see in dry soil.

1. Identify within your property the most consistent and largest mapped soil type.

2. Go to the center of the largest units and use a soil probe or shovel to expose a profile to at least 18 inches depth.

3. For each exposed profile, measure the depth of each layer or soil horizon, and draw the corresponding layers on the soil data form. Match what you see to the detailed description for that soil type found in the county soil survey manual, which you should take with you to the field.

4. For each layer, using the same form, complete the descriptive information, defining major differences in depth, texture, and color.

5. Use the descriptions in box 1.4 to characterize the soil texture (see figs. 1.8 and 1.9).

6. From the surface layer, collect a quart-size zip-lock plastic bag of soils, labeled with either the GPS point number coordinates or a reasonably accurate diagram of the sampling location, and the soil type as identified from the county soil survey.

7. Find two or more mapped edges or boundaries of these larger mapped soil types and repeat this same examination. If edges do not correspond to the soil map, make adjustments on your basemap. Pay particular attention to changes from what you found near the center of the soil units. Does the soil still meet the description for the major types? What changes are present? Compare the soil type with the adjoining type. Are changes intermediate?

8. Repeat the same two-step process with other common soil types on the property. Record and collect samples as before. Particularly note small landform features that were likely ignored in the NRCS mapping. Can you match the soil type in them to one described in the country soil survey manual?

9. Review your soil map and land cover map to identify any additional locations that also need to be sampled, and continue the process as your time permits.

Based on your interpretation, modify the basemap for your project site. Review task 4 and consider whether you have additional insight to the reasons or conditions for the distribution of cover types or land use. For example, you may find some areas dense with invasive plants. Perhaps you will find heavily compacted soils there. You may also discover additional wetlands that have been disguised by disturbance. Perhaps soil profiles reveal places where topsoil is gone, exposing clay-rich subsoil, or places where the topsoil from an adjacent upland eroded and accumulated over an existing soil. This was discovered by the authors on a restoration project in Costa Rica (see fig. 1.10).

The samples you have collected can now be analyzed, or dried and archived for later use. Do

---

BOX 1.4. FIELD DETERMINATION OF SOIL TEXTURE AND STRUCTURE

Take a small handful of soil from the layer you wish to characterize. Wet it so that you can squeeze it into a ball, like cookie dough, but not so much that water drips from it. Now squeeze the mud between your thumb and forefinger, attempting to flatten it into a ribbon as long as possible. Don't worry about thickness as you extrude the soil between thumb and finger. Can you feel grit? If so, you have a sandy textured soil. Loams will ribbon to an inch or so, and clay soils to well over an inch. Sandy soils may not ribbon at all. A sandy-loam will feel somewhat gritty as you ribbon it to half an inch or so. A loamy clay or clay-loam will ribbon more without feeling gritty, the latter somewhat more than the former. A sandy-clay or sandy-clay-loam will ribbon more, the latter less than the former. (See fig. 1.8.)

Soil structure (fig. 1.9) is determined by exposing a relatively undisturbed face of the profile, or lifting a shovel of soil if you are only interested in the surface few inches. Use a pen knife to gently probe and tease the soil apart. It is the shape and size of the structural units you will want to observe. One of the most common structures for topsoil with a reasonable amount of organic matter is granular. Granular peds (individual units) tend to be more or less rounded, up to a quarter of an inch or so in diameter. Very sandy soils may have no structure, with individual grains falling apart. Especially in subsoil, watch for blocky structures, with units of varying size up to two or three inches. If longer than wide, they may be columnar. Sometimes, peds are prismatic. In a few instances soil may be platy with peds flattened. Soil structure is also rated as structureless, weak, moderate, or strong, according to how much integrity the peds display. Weak structure is very easily broken apart. Strong structure resists being broken. An excellent reference for this and additional information is available online from the National Aeronautics and Space Administration (NASA) at http://soil.gfsc.nasa.gov/pvg/prop1.htm.

---

not dry the samples, however, if there are locations where you want to know the types of plants likely to germinate from the soil seedbanks. To do this, follow these three steps:

1. Spread samples of approximately one cup of soil evenly over trays or flower pots of sterile sand, available from garden stores. Then place the pots or trays in a greenhouse or some protected location where you can maintain temperature at 75 degrees or higher.

2. Water the samples daily and watch for germination. Juvenile plants are difficult to identify without considerable experience, so allow seedlings to keep growing. As they mature, you will need to water more frequently.

3. Once you can identify a plant, remove all plants of that species from your samples, disturbing the soil as little as possible. Record the numbers of individuals. You may find germination continuing over many weeks. Discontinue only when you have no new plants after two weeks.

Each square foot of top soil to a depth of eight inches weighs on the order of 45 pounds. The one-cup sample of soil represents about half a pound. Thus, you can get a rough estimate of the number of viable propagules per square foot for

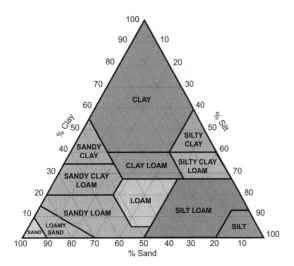

FIGURE 1.8. A soil textural triangle

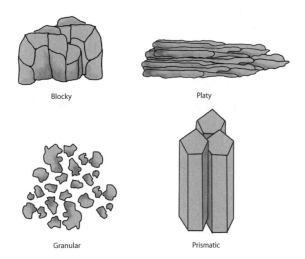

FIGURE 1.9. Examples of structural peds found in soil

each species by multiplying the number that grew from each sample times 90.

This seedbank analysis reveals whether desirable native plant species are present and able to germinate. Especially on former agricultural land, most of the species will be nonnative weeds.

Watch especially for noxious weeds that may present problems during restoration. On one very large (thousands of acres) restoration project, we saved $50,000 in seed purchases by relying on the seedbank.

Soil samples you collected can also be used to investigate the chemistry of soils. If soils were taken from previously cultivated fields or orchards, where they receive runoff from neighboring farms or manufacturing facilities, you may want to test pH, nitrogen, phosphorous, or for presences of pesticides. Your county extension office will have sample bags and instructions for preparing samples. The samples are sent to a state laboratory for analysis. Results will be returned with directions for interpreting data. If your concern is with salinity or potential toxic chemicals, be sure to indicate that with your samples. Testing for toxic chemicals is expensive. Consider sequential analysis. Initially test only soil from areas of greatest concern. If preliminary tests are positive, you can always submit additional samples.

Air dry soils you plan to hold for future testing for one to two weeks, depending on humidity, in a protected location. They should be powdery dry. These can used for additional analysis or to test for changes over time. Label them well, box them, and store them in a cool, dry location.

Surface Hydrology

In most landscapes, direct and indirect effects people have had on soils and vegetation have altered the way water interacts with land. Restoration often is focused on restoring surface and shallow groundwater hydrology. Often, surface water infiltrates into the most permeable soils in uplands and some percentage will reappear as springs and seeps near the bottom of slopes, usu-

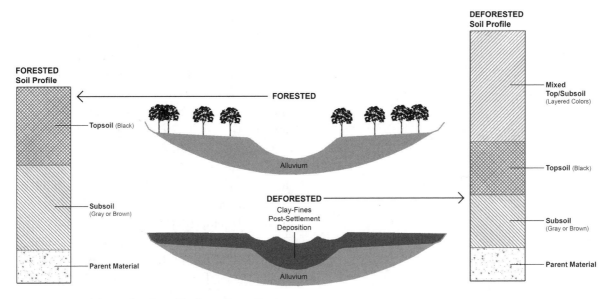

FIGURE 1.10. A buried soil profile from Costa Rica

ally along the margins of streams or wetlands. Restoration of shallow groundwater hydrology is often accomplished by removing artificial drainage systems and restoring plant communities. Native vegetation encourages infiltration in even the heaviest of clay soils. We briefly describe techniques for monitoring hydrology later in step 6.

Task 6. Map Drainage

The immediate focus is understanding existing drainage features. As with the NRCS soil map, topographic maps are not finely resolved, and often show only the largest of the drainageways. For example, the contour interval on the USGS topographic map is rarely less than ten feet in vertical relief, much more in mountainous regions. Any drainageway with less relief will likely not be included in the map. Because every spot on the land is part of a drainage network (see fig. 1.11), it

is important to map this system on your land. On very large projects, the USGS mapping may give you enough information in all but a few locations, such as where you need to alter a drainage channel, or do stream or wetland restoration. For these types of projects you will need more detail, requiring additional equipment (see appendix 2) to take measurements on the channel and determine tributary acreages.

Drainageway and tributary watershed mapping can be accurately done with USGS topographic maps supplemented with additional field review to identify smaller watersheds and drainageways. The goals of this task include the following:

- Understand the drainage network on the land and how it aligns with soil types.
- Identify and map smaller features not noted on the topographic map and correct any inconsistencies.

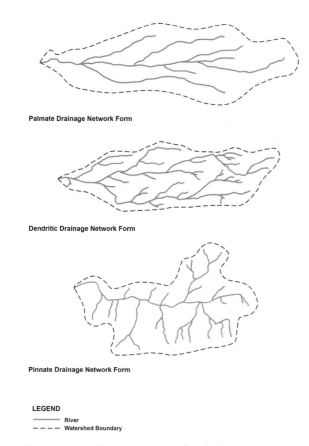

Palmate Drainage Network Form

Dendritic Drainage Network Form

Pinnate Drainage Network Form

LEGEND
———— River
– – – – Watershed Boundary

FIGURE 1.11. Drainage networks of tributary watersheds

- Map erosion and depositional areas associated with drainageways.

This process involves using the basemap, again making direct annotations on a copy or using an overlay of Mylar film.

1. Use the basemap with the USGS topographic map and identify drainageways and topographic divides that separate tributary watershed areas for each. Use the stream channel and drainageways mapping data form (data form 1.10, appendix 1), and complete for each drainageway the basic descriptions or measurements.

2. As you walk the land, identify and map any additional drainageways where there is at least periodic water flow. Record the movement of stormwater runoff over the land. A hand-held GPS unit is very useful for this. Walk the centerlines of drainageways recording points (way stations) as you go. These can later be downloaded to your map. If drainageways have perennial flows (these usually are mapped by USGS), use the symbol for a stream. If they are intermittent, use the intermittent mapping symbol. If they simply are overland flow, active only during storms or snowmelt periods, the areas affected should be noted. It is also useful to walk perpendicular to these in several locations and map the "height of land," where you leave one watershed and begin descending into the next.

3. If you use a handheld GPS, make sure you also record benchmark locations on the property, so that the GPS locations can be registered when you plot them on paper. We usually use the centerline of a roadway, fence corners, or other permanent features each time we use a hand-held GPS.

4. For any drainageways that have been channelized (straightened), denote those reaches, and distinguish them from the nonstraightened reaches.

5. Now return to representative channelized and unchannelized (reference area) sections, particularly those with likely or known perennial water flow, and take measurements to complete the stream channel and drainageways mapping data form (see data form 1.10, appendix 1). Some of the measurements taken at this time will be useful for restoration planning, especially if stream restoration or stabilization is a potential part of the plan.

6. As you are measuring ditches and stream channels, keep your eyes open for farm tile lines.

These almost always end and discharge into ditches and stream channels. The ends of the tiles often are well concealed in dense vegetation on the banks, making them difficult to spot. Often the presence of vitreous or reddish-brown clay tile chips, or broken tiles that have fallen from the bank, will provide a clue. Tiles will have been installed most commonly where the grade of the land adjoining the stream or ditch is relatively level and where soils are poorly drained clay, peat, or muck types.

7. Find the outlet ends of the tiles and map them. If you find two adjacent tiles, measure the distance between them. Spacing was critical to drain soil effectively and tile lines usually were installed uniformly over the area. Once you have figured out the spacing, you can measure from known tile lines and find others. Most tile lines were 15 to 50 feet apart. Tiles are important to decommission, especially for wetland restoration. You will need to plug tile outlets, or break them up with a subsoiler to restore wetlands. A subsoiler is a heavy blade pulled behind a tractor that knifes up to six feet into the soil. Tiles will soon plug where a subsoiler breaks tile lines.

Invasive Species

One of the most important restoration activities is the control of nonnative species. Sometimes there also is a need to control native invasive species. In some cases, nonnative invasive species are closely related to and similar to desirable native species. As suggested by the name, invasive species invade habitats, usually aggressively. Invasive species tend to be nonnative, not part of the native community, aggressive and persistent, and tend to occur in a wide range of ecological conditions.

Invasive species are becoming more common throughout the world. In some cases, only a nonnative genetic strain of a species shows invasive tendencies while native populations are not a problem. An example is giant reed grass (*Phragmites communis*). For years it was not known if it was a native or introduced. We now know there are both strains with important differences, and these genotypes have hybridized allowing the genes causing invasiveness to enter our native population.

Invasive species generally behave differently from native species, demonstrating "weedy" characteristics. Following their initial establishment, they often demonstrate these characteristics:

- they remain quiescent for some time, then rapidly increase in abundance;
- they form dense growths or monocultures;
- they spill over into other communities from initial colonies;
- they crowd out other vegetation;
- they are widely adaptable to a range of habitats;
- they are less attractive to native wildlife species;
- they follow zones of perturbation or disturbance (e.g., erosion/sedimentation);
- they, at least initially, are more common along roadways or trails, suggesting modes of introduction;
- they are more aggressive in nutrient-enrichment zones, such as manure runoff areas;
- they are more aggressive in areas where native species are disadvantaged, such as areas of soil compaction and salt deposition.

For these reasons, it is important to recognize invasive species early, certainly at the beginning of a restoration if they are already present.

Task 7. Map Locations of Significant Populations of Invasive Species

Recognizing invasive species is not always easy. Every state has listings of invasive species, often with county distribution maps, and websites devoted to them, including recognition and control. Many county extension offices will also have information on them. We recommend that you download information for species of potential concern in your county and learn to recognize them, or stop at the extension office and get fliers or other information available.

The objectives of this task include identifying and mapping invasive species in reference to the cover map, using symbols to indicate relative abundance (see fig. 1.12 as example); and developing an understanding of invasive species in your project, especially as they may relate to soil compaction, erosion and deposition, and nutrient enrichment.

Achieving the goals of this task follows a similar field-mapping process as previous tasks. Additional equipment will be needed (see appendix 2) to identify and collect specimens of plants you believe to be invasive species.

1. Before beginning fieldwork, obtain the state or county listing of invasive plant species that are known to be aggressive and persistent in your area. You may need to parse from the full list the species that are of special concern. Spend as

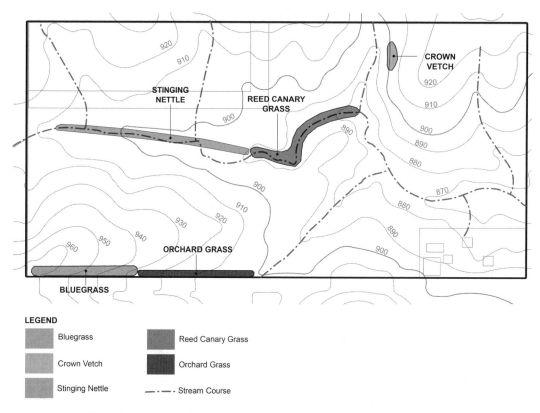

FIGURE 1.12. Map of invasive plant species, Stone Prairie Farm, WI

much time as you need to familiarize yourself with the key species so that you can recognize them in the field.

2. During the growing season, ideally from early summer through early fall of the year, revisit areas of disturbance, nutrient enrichment, and erosion and sedimentation zones. Especially check around farmsteads, roadways, and similar types of settings to locate invasive species.

3. When you locate an invasive species, usually a patch, record the location, or mark it on a copy of the basemap.

4. At each location where you find invasive species, complete the invasive plant species data form (data form 1.11, appendix 1). This involves estimating relative abundance of the plants. If possible, record the GPS location of invasive species on the data form.

5. Collect a representative plant specimen for positive identification. When possible, obtain specimens with leaves, stems, flowers or fruit, and roots, for nonwoody plants. Place collected specimens in plastic bags, neatly folded to minimize damage to the specimen, and record location and sample number.

6. We recommend that you make a record of the plant specimens in two ways. First, spread the plant on a plate of a photocopy machine, then place a clean white page on top of the specimen, and obtain a clear image. Second, position the specimens in folded newspaper sheets and place the specimens in the plant press until dry.

7. Confirm identifications from reference books or ask local experts to help. You may be able to scan and email the shadow images to some experts who can give you identifications based on the images. For uncommon species, experts may need to see pressed specimens. Check with your county Extension or NRCS office for assistance in identification and control.

8. Cross-reference your field interpretation, data form records, and map of the species and relative abundance of invasive species to correspond to specimen identifications.

Many invasive species will not need management during your restoration program. They simply will fade out as desirable native vegetation becomes established. The persistent species, the well-known problematic invasive species, may require management during and after restoration.

Assessing Stressors

The last task in evaluating your land primarily comes from summarizing and synthesizing what you have learned. This synthesis is captured with a summary map of the *stressors* that you believe have affected, or continue to affect, the physical, chemical, or biological environment on your property. You already have the clues in your maps of soils, vegetation, hydrology, and invasive species.

Stressors are chronic (e.g., effects of erosion or nutrient enrichment) or stochastic (e.g., a flood that damages a stream channel or floodplain forest) effects that operate at different scales of time and space. Stressor mapping is an important way to summarize what you have learned from the work completed so far, and to identify stressors that you likely will need to address with your restoration plan.

Task 8. Assess and Map Stressors

The goals of this exercise are twofold: (1) Create a summary map of existing and past disturbances such as erosion and deposition of soil, nutrient enrichment, salt and other contaminants, soil compaction, and so forth. (2) Map the stressors that you can control, those you cannot control, and offsite stressors you are not likely to be able to control or mitigate without a successful partnership with neighbors. See figure 1.3 as an illustrative example.

The process is straightforward:

1. Create an overlay of the basemap of features such as erosion, sedimentation, nutrient enrichments, eroded and channelized drainageways, and locations with anomalous soil horizons such as where sediments have buried existing soils. Note either symbolically or in your notations those that can be addressed within your property boundary and those that you cannot address, at least without neighboring cooperation.

2. Add your map of invasive plant populations.

3. Either with the same map or a summary map, draw boundaries or circles around stressor zones on your property. Make sure you have a good description of each.

4. Create a new map or Mylar overlay labeled "future potential stressors." On this, begin a process of inquiry. Challenge yourself to list, map, and describe potential and likely future stressors that may become a threat to your restoration. Typically, these result from changes in land use (e.g., ranches stepping up the number of grazing livestock). For nearly all regions, you can expect invasive species to become more problematic, but identify areas where they will most likely occur.

Stressor analyses are very important as you move forward with your restoration program.

Continue to seek new insights into stressors that affect your land. As an ongoing assignment, you should continuously adjust your summary stressor map and the future stressor map for your property.

Summary

You now have existing condition maps for land use and land cover, soil and surface hydrology maps, an invasive species map, and a map of the stressors affecting your restoration. You will also have some good insight into understanding the drivers that are influencing the land, and how invasive species have responded to various stressors operating now and in the recent past.

With this information, and from having gone through the process, you are ready to investigate the historic changes, and past and present conditions of your project landscape. In the next step we will explore this history and its relationships with what you have learned in this step.

Notes

1. Neal S. Eash, Cary J. Green, Aga Razvi, and William F. Bennett, *Soil Science Simplified. 5th ed.* (Ames, IA: Blackwell Publishing, 2008). *This book has been reprinted and improved through thirty years of use as an introductory text. It is wonderfully illustrated and written for readers with little or no background in soil science.*

2. Soil Survey Division Staff, *Soil Survey Manual.* (Washington DC: Soil Conservation Service, U.S. Department of Agriculture Handbook 18, 1993).

Investigate Historic Conditions

If you want to understand today, you have to search yesterday.

Pearl Buck

Like an organism, ecosystems have the capacity for continual renewal but often require some intervention to assist the process. The primary aim of ecological restoration is to restore the land's capacity for renewal. Before we can determine proper treatments we must understand the land, how it looked and functioned before its health was impaired. We also need to understand how the health of the land is continuing to be negatively impacted. In this step, we walk you through the techniques and potential sources of information that provide insights into how your land once looked and functioned.

The characteristics of ecosystems—structure, composition, and ecological functions—are less familiar to most of us than human physiology and behavior. Still, just as you do not have to be a sociologist or psychologist to understand much about yourself and your relatives, you need not be an ecologist to gain much understanding of how historic conditions have influenced the present state of your land and the ecosystems on it. The present condition of the land reflects its history and the forces that continue to shape it. This was beautifully illustrated by May Theilgaard Watts, with examples from across North America, in her classic

book titled *Reading the Landscape*.[1] We recommend her book as a good introduction to this phase of planning your restoration.

Understanding the Historic Landscape

We explore history as a process in planning ecological restoration for the following reasons:

- To help explain what you found during the "existing conditions survey and mapping" exercise.
- To better understand reasons for the changes in the land.
- To explore what can be restored and what is not likely restorable, at least now.

Unlike peeling an onion, revealing one layer after another, learning about your land is usually not a systematic, sequential process. With ecosystems, you may discover layers (bits of information) in any sequence. This is more like looking at a circus through a hole in the tent. With little concept of what a circus is, at one moment you see a clown, another you see a couple on a trapeze, or a horse galloping around with a near-naked lady on board. Looking in the peephole of the tent, whether you see the horses or the elephants will seem random. These glimpses will make little sense at first, until you have studied enough to begin to understand the bigger picture.

Our formal education teaches us to look for linear relationships. In school, the lessons and chapters in our books introduced the most basic knowledge first and then layered on more and more details and understandings in subsequent chapters or classes. The linear equivalent of learning about your land would begin, perhaps, two hundred years ago and sequentially cover the land's physical, chemical, biological, and spatial conditions, and how they changed up to the present. However, few if any relationships in nature are linear. Indeed, if a linear relationship is found, the sharp ecologist will immediately look for an explanation.

Learning about land is further complicated because many pieces of the grand puzzle are now missing. This requires that you develop working hypotheses to explain what may be missing, and over time, test your guesses as to how the pieces fit together. Imagine putting a jigsaw puzzle together without the picture on the box as an aid and with many missing pieces. You begin with only a vague idea, but as pieces come together, you begin to guess with greater and greater accuracy about the bigger picture. The more the pieces come together, the easier it is to fill in gaps.

In this process, we note facts or events, link them with hypotheses, test the timing and understanding of the events and facts, and derive relationships and insights. As with the jigsaw puzzle or the circus, the tidbits of insight will mean little at first, but will soon begin to tell a more complete story. Herein lies the excitement associated with exploring the history of your land. You are on a discovery trip with, at best, a vague idea of what you will discover and when.

As we explore the history of the land and begin to assemble our ideas about restoration, we should keep the following in mind:

- While we may learn much from the past, we cannot recreate the past through a restoration process. Not only have prevailing conditions changed, we never can completely understand the past. By focusing on some date-specific or condition-specific historic point, and designing the restoration around that, we lock in on a static goal. "Stasis" does not exist in nature.

- Our primary reason for understanding history is to learn how the land appeared before humans altered it, and what the land might become with restoration and management. Once we understand the land, we can better jump-start the drivers to set in motion the ecological processes that lead to an alternative trajectory from the one the land is now on. By trajectory, we mean the succession of communities that occupy the land while they themselves continue to change and morph, under the influence of management and changing physical, biological, chemical conditions. In later steps we will focus on designing goals and objectives, and the design and implementation of a monitoring program to help you better understand and track trajectories.

- Learning about and communicating the history of your land is useful not only in strengthening your plans, but can also be one of more powerful ways to engage others—friends, family, neighbors, colleagues—in the restoration process. Successful restorations require that you directly or indirectly work with others. Being able to paint the history of your land is like describing the circus to those you want to encourage to attend with you. It helps others to see and understand your intentions and goals.

Learning about the Past

The well of history is deep. It is as rich as the time you have available to explore it. The primary tools for doing so are the answers to key questions. You will need to loosely allocate time for gathering information at the beginning of your planning, based on its value and accessibility, but keep coming back to this investigation over the years. Always pick the "low-hanging fruit" first, but some valuable information may require considerable effort. The quest is a process not unlike creating a painting. The first layer on the canvas should provide a durable foundation for the subsequent coats and colors so they survive time and do not bleed through, or fade away. The first layer(s) also must provide the structure to allow for the richness of the future shadowing, transparency, and depth to become apparent as layers are added and as the painting ages.

The questions that will frame your thinking should lead you to an understanding of the physical, biological, and temporal (history). Your inquiries into history work best if you organize historic information under these categories. Over time, perhaps years or decades, you will find additional information to address questions that will continually arise and reach greater insight and resolution on some earlier questions. You also will undoubtedly come up with many new questions. Some will never be answered.

The relationships between physical and chemical (e.g., soils, geology, water), biotic (communities of microorganisms, plants, and animals), and time (e.g., evolution of species, soil development, and successional changes in ecosystems) are very complex, but we can tease out important insights.

Key Questions

There are questions that provide important clues to the past. Remember that when investigating complex questions, information does not come, necessarily, in any particular order. Answers to the important questions may not be understood if addressed at the beginning of the inquiry. Often, small pieces of information incrementally lead to answers, and you must be patient and persistent in piecing together information. This also means that you have to avoid prematurely locking onto interpretations; remain open to new information, even for years. Restoration is a continual learning process.

Task 9. Complete Historic Conditions Data Form

To get you started, we provide a list of questions (boxes 2.1, 2.2, 2.3) and a historic conditions checklist (box 2.4), which will help you focus on the kinds of information you should obtain from the land you want to restore. These are not intended to be all inclusive; you may have other questions now or later. Review the historic conditions data form (data form 2.1, appendix 1). Use this to record any answers you have as we go through the questions, but you will need to come back later to fill it out more completely.

BOX 2.1. QUESTIONS FOR UNDERSTANDING THE PHYSICAL SETTING

1. What geological features are prominent on your land?

2. What different types of landforms are present?

3. Are the surface substrates transported from other locations, or have they developed in place from the breakdown of bedrock?

4. How have water, wind, and the interplay between them altered the land?

    a. Have the rates of movement changed? For example, erosion processes accelerate under more humid conditions. Vegetation stabilizes soil.

    b. Are there scars and patterns suggesting historic water movement?

    c. Does it look like human modifications of historic patterns of the landform or vegetation have occurred on your land?

    d. Are there sources of water on your property that may influence the existing conditions? For example, there may be springs that emerge from the toe of a bluff where white deposits of minerals are found.

    e. Have the water courses been modified, such as by damming, dredging, or channelization in or around your property?

BOX 2.2. QUESTIONS FOR UNDERSTANDING THE BIOLOGICAL SETTING

1. Are there recognized "preserves" on your land or in your neighborhood known to contain remnants of historic vegetation, such as prairies, wetlands, forests, and deserts?

2. If you live in farm country, are there areas that are so dry, rocky, or inaccessible that they have not been plowed or heavily grazed by livestock?

3. Are there stories, even myths, told by elderly neighbors about the last wolf or elk in your township? How about fishing stories about huge catches?

4. Are there stories about other wild harvests (fruits, nuts, mushrooms, herbs)?

5. Do stories still exist that tell of annual animal migrations?

6. Is your land vegetated by invasive, nonnative grasses, herbs, shrubs, and trees?

7. Is the barn or house on your land made from local timbers and wood?

Sources for Exploring the History of Your Land

There are many sources for exploring land history; we discuss only a few of the most important but include a list of some other potential sources that may be available for your land. Depending on where your land is located, relative to urban areas or well-studied lands such as national parks, there may be much historic documentation and analysis already done. More commonly, you will have to hunt for all but the most basic history.

BOX 2.3. QUESTIONS FOR UNDERSTANDING THE HISTORICAL SETTING

1. One way to learn about the history of your land is to determine the ages of trees growing on the land. Have you counted the growth rings in any trees?

2. Do stories exist in your neighborhood that describe how your property looked one hundred or more years ago? What species were present then that no longer occur?

3. Is there a difference in the species and size of trees on your land compared to nearby land? If so, why?

4. What do you know about the former owners?

5. Is there a cemetery on or near your land? Are there neighbors who can tell you about the people buried there?

Reference areas. Reference natural areas are one of the best ways to inform you of what your land was like before people altered it. An important step in preparing for your restoration planning is *reference area analysis.* This process involves finding, measuring, and correlating nearby natural areas to your restoration site. Sampling includes measuring the composition and structure of the plant communities and making species lists, but you also may want to probe soils to make sure they match your project reasonably well. The most important reference areas are areas of land, water, or wetlands that most closely resemble and function like the historic settings in your project. Many states have natural area inventories and you can access existing data collected by scientists. Many natural areas, both public and private, have become the focus of preservation and protection by land trusts, departments of natural resources, organizations such as The Nature Conservancy, or even local community groups. Make arrangements to visit and make additional measurements or sample vegetation in nearby reference areas that are similar to your project site. On public properties, a scientific investigation permit may be required. Some states may

even allow modest seed collection if you use it for your local restoration. This ensures that species being introduced are from local genetic stocks. In natural areas that are privately owned, landowner permission is needed for any sampling or seed collection.

Finding remnants not previously reported by a natural area inventory can be a fun process. Much like a treasure hunt, Steve and his family wandered the southern Wisconsin landscape looking for remnants on neighboring farms. They found dozens of small patches of native prairie, wetlands, savannas, high-quality forests, and small streams that were not identified by the natural areas inventory program. Once landowner permission was obtained, these became reference areas from which data (e.g., soil characterizations, species lists by plant community, hydrology data, and some wildlife data) were collected, which was used to refine the restoration plans for their farm. Use aerial photographs and topographic and soils maps to spot likely places, and correlate slope, aspect, soil types, and hydrologic conditions with landscape positions in your project. Then match species and other information in the planning process. The monitoring

---

BOX 2.4. HISTORIC CONDITIONS CHECKLIST

- Physical conditions
  - _____ roads (paved and unpaved), driveways, rail lines, human and animal trails
  - _____ quarries and other excavations
  - _____ homes, farms, outbuildings, permanent equipment (pumps, wells, pipes, etc.)
  - _____ fence lines, tree rows, and other linear features
  - _____ hydrologic features (streams, ponds, springs, seeps, etc.)
  - _____ wet ground (wetlands, irrigation lands, septic systems, farm runoff areas)
  - _____ drainageways (erosion features), drainage infrastructure (tile lines, bridges, fords, etc.)
  - _____ bare or exposed soils, utility scars (construction sites, bulldozing, plowed fields, etc.)
  - _____ archeological resources (grave sites, council ring sites, bone, buffalo jumps, arrowheads, etc.)
- Biological Features
  - _____ crop patterns, crop land soil preparation patterns, and harvesting patterns
  - _____ trees, shrubs, and coarse woody debris
  - _____ partially cleared lands of native grassland, wetland, brushlands, or forest cover
  - _____ water condition: moving water vs. pooled water, temporary vs. perennial water
  - _____ vegetation cover types (wetlands, shrub-carr, forest, sedge meadow, etc.)
  - _____ chaining, bulldozing, root chopping, roller chopping, plucking, patterns of woody vegetation management
  - _____ animals: livestock and wildlife numbers and trail density
- Time
  - _____ burn scars showing recent blackened areas or responding vegetation patterns and stands with uniform textures and patterns
  - _____ blowdown scars, ice damage to trees, tornado or hurricane tracks
  - _____ changes in any of the checklist items between photographs of different dates
  - _____ deforestation, plowing of land, conversion of wetlands, drainage features, etc.
  - _____ paths, trails, and roads between neighboring properties that may suggest relationships
  - _____ property boundary changes, including fence-line relocations, etc.

---

methods reviewed in step 6 are the same methods that can be used for gathering the basic information needed for restoration planning on your project (see fig. 2.1).

Maps and descriptions of the historic vegetation and ecosystems over most of the United States have been completed to various levels of resolution (see fig. 2.2). In addition, there are books with plant and animal species lists associated with many, if not all, of the major ecosystems included in these maps. For general background, it may be useful to review this level of information as you zero in on specific community types in your project and reference natural areas.

FIGURE 2.1. Example map of reference area features and relationship with Stone Prairie Farm, WI

Aerial photographs. "Photographs don't lie," or so the saying goes. But, unlike a book that may provide an interpretation of some event or condition, aerial photographs do not come with this interpretation done for you. Photographs provide a snapshot of the condition(s) of the land at the time the photograph was taken. If the resolution of the photograph is high enough, you may be able to enlarge it sufficiently to distinguish individual plant communities, the quality of the vegetation on stream banks, and what crops were being grown. You can learn about the condition and changes in neighboring properties and consider changes in the larger context. By comparing photographs taken at different times, you can follow changes on the land. Use the earliest and most recent photos available with at least some in

between when available. Ideally, we would like to see conditions evaluated every decade or so.

For Steve's farm in Wisconsin, aerial photographs were available every seven to ten years starting in 1937. The early photographs were extraordinarily high-quality military images and could be enlarged to see great detail. Images for some of the subsequent years were not as good, and blurred quickly with minor enlargement. In the late 1970s, color images became available. Still later, color-infrared images became available. Since the late 1980s, satellite images have been available as often as every nine to twenty days over every location on earth. The early satellite images are too low a resolution to be of much use, whereas recent technologies are digitally available (often online) with resolutions of a few

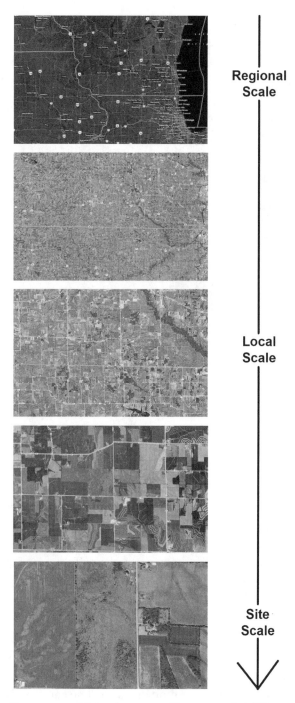

Regional Scale

Local Scale

Site Scale

FIGURE 2.2. Natural communities mapped at different scales

feet (e.g., one meter). Resolution this high means that you should be able to see most features on the ground such as trees and fences, and distinguish differences between features that are at least a few feet apart.

It is best to use photographs at a common scale, with enlargement or reduction as necessary. If you are asking a federal agency or the National Archives (www://archives.gov/research /order/maps.html) to reproduce photographs from their files, ask for the scale you desire so that you need not enlarge or reduce them. Imagery, even old paper photographs, can be scanned and digitized. Once in a digital format, with software such as Photoshop, you can easily revise the scales to suit your needs.

The interpretation of aerial photographs or satellite images is essentially a mapping exercise. The simplest process employs onion-skin tracing paper from an art supply store and nondestructive sticky tabs or drafting tape. Securely attach the tracing paper over the photograph along one edge. By leaving the rest free, you can lift the tracing paper if needed to see the unobstructed photograph to confirm images. If you can get clear Mylar plastic film instead of tracing paper, you see through it to the underlying photograph. Otherwise, you will need to find or rig a light table. This is simply a large box covered with plexiglass with a bright light source that shines through the photograph, allowing you to see the image even through tracing paper. Before we owned a light table, we taped the photograph and tracing paper to an exterior window on a bright day. Although not as convenient because of the vertical orientation, the result was equally good.

1. Mark at least three corners or other features evident on the underlying photograph on the

tracing paper or Mylar. This allows you to remove and replace the tracing back in the exact same position over the photograph if you need to come back to adjust something later. Once registered, the work begins.

2. Using the historic conditions checklist (box 2.4), go down the checklist of physical features and view each of the aerial photographs to see what you can learn and what seems important enough to interpret and trace. Start with the roads, driveways, rail lines, trails, buildings, and other anthropogenic features, then move to land use (tillage, forage, pasture, etc.) and land cover (forest, grassland, crops, etc.), on each photograph. This process starts revealing quickly the broad-brush changes in your land and the neighboring properties. As you go back to look at each photograph, begin to look for greater detail. A hand lens may be useful to take a closer look, especially if you are working with paper copies of the photographs. If you have everything in digital form on computer, you can pan around and then enlarge or scale back to find and explore and interpret the land.

Give each feature you map a different symbol, and attempt to standardize with the symbols used by the United States Geological Survey (USGS). Denote the differences between various stream courses (e.g., perennial streams versus intermittent streams) and water bodies (human-made ponds versus natural ponds) that can easily be depicted with the standard map symbols. You will quickly see differences in conifer plantations and natural conifer stands, young hardwood forests and mature forests, even sedge meadows and pasture.

You may find in developing your interpretation, that the tracing paper will become very busy, leading to confusion, and affecting your ability to remember what it all means. We find it useful to do multiple overlays. For example, you might do physical features, biological features, and temporal changes separately. Typically, we do the linear features on one layer, then place and register a second trace over the former (leaving the previous attached over the photographs) until we have completely gone through all features we want to map. This sometimes includes three or four layers. Often, new features not anticipated in the catalog need to be defined and mapped with a new symbol. In fact, most of the catalog features will require that you make up and record the definition of the symbol(s) you have used in interpreting your land.

3. After you have the layers traced for a photograph, summarize conditions by retracing the layers onto a single sheet. With a good symbolic marking strategy (symbols should require little room and not obscure other information on the trace), we can summarize all information from each photograph on a single new trace sheet. Doing this sets you up for the temporal analysis.

4. Understanding changes over time is accomplished by taking the summary tracings from each photo year and superimposing them on the light table, on clean tracing paper. You may need to do this in steps, comparing two tracings at a time, although it may not be necessary to analyze each date. Determine the key changes, and document a timeline or chronology that refers to the trace overlays. Then, summarize them to tell the story of the changes that have occurred over the land. In *Restoring Ecological Health to Your Land*, we provided an example of this summary for Steve's Wisconsin farm. When were fields cleared, tiles installed, ditches dug, streams straightened, or pastures converted to row-crops? If contour tillage was used, when did it begin?

5. Ideally, you should finalize a record that you can share with others. To do this, scan each historic photograph to make a digital record. If you do not have a scanner, check with computer-savvy friends. If they do not have one, chances are they know where you can get it done. The digital images, including the summary image, can be presented sequentially to tell a story about the land. This has been the most powerful way for us to share with others what we have learned. Neighbors, especially those who have been on their land a generation or two, will be especially interested and will add many anecdotes and details to your story, leading to greater understanding of what actually happened on your land or theirs.

Historic Maps, Documents, and Other Sources of Information

Many types of historic maps and written documents exist that may paint additional layers of richness on the underlying understanding you have of your property. These resources differ in quality, coverage, and usefulness from document to document and area to area. From each source of information, your goal should be to transfer information into a format that can be spatially referenced using either a digital form or the same tracing paper process. If the information applies broadly, for example a historic drought over the region 120 years ago, at least note this in your journal. If it applies to a portion of your land, for example, a fire swept across the south half of section 24 in May 1913, make a note on your summary map as well as in your journal.

Original land surveys. Sanctioned by the General Land Office, at the direction of President Jef-

ferson, this was a survey process that laid out the sections and townships and allowed for the current legal descriptions used in the United States on the title or deed to your land. These surveys were completed by hired surveyors dragging metal measuring chains across most of the country, aligned on north-south and east-west compass bearings. At every quarter-mile (quarter-section), half-mile, and mile (section corners), they erected survey markers to allow the corners to be relocated. They also both mapped and wrote descriptions of each major vegetation type they encountered as they measured their way across America (see example of original land survey notes and plats, figs. 2.3 and 2.4). Their measurements and vegetation descriptions have been used to create historic vegetation maps for the time of survey. With current soil and topographic maps, ecologists and cartographers were able to extrapolate the original survey vegetation types across the land and adjust the boundaries in locations where the original surveyor was vague or where it was not mapped. These maps are available in digital form from most state historic societies, county courthouses, and often through state government land records offices. In some states, these records are still available only on microfilm. You should obtain the summary plats for your township, any river or waterway plat maps, and the original surveyor journal records. Spend some time capturing the images in your mind and describe in your journal how the surveyors saw your land and the nearby land. Note the dates of the surveys.

Oral history interviews. Older people in your neighborhood are often a great source of information. Develop and ask questions to start the conversations, and consider recording the interviews. Often questions lead to remembering old

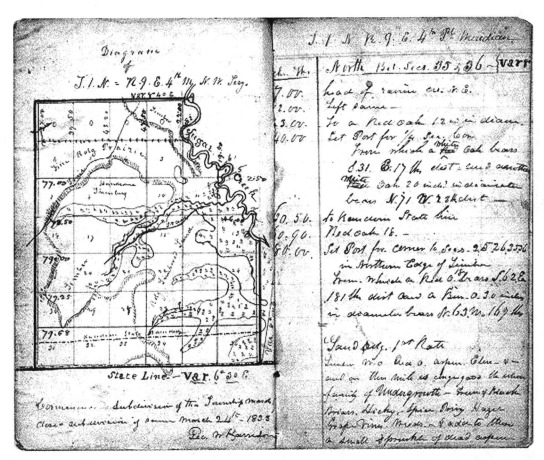

FIGURE 2.3. Example of original land survey notes

photographs or letters that contain insights into the past. Use the accumulating information to frame successive interviews with the same or other persons. Steve learned about the last passenger pigeon hunt in the neighborhood, and when food was scarce and meat was needed, circle-hunts when neighbors would walk the land and drive game toward the center. There were stories of one-room school houses; trap-lines to raise money for the family needs; the final days of the wolf trappers and the bounties they received. Use the oral history interview data form (data form 2.2, appendix 1) to help organize this infor-

mation. You can apply any of the framing questions covered earlier to develop your interview questions. Create a numbered list of questions and then place the number of each question on a local and regional map such as a Gazetteer, so that you can geographically locate the place reference by the persons you interview. Along with the mapped identification number, it is useful to embed the initials of the person's name whom you are interviewing.

Tree-ring analysis. Tree-ring analysis, or dendrochronology, uses the pattern of growth-ring

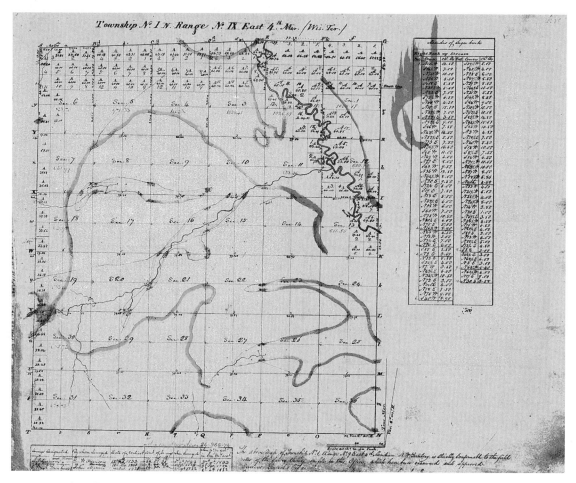

FIGURE 2.4. Plat of a survey map

size in old trees to gain insights into past climate. Wide rings suggest plentiful moisture while narrow rings indicate periods of drought. The process begins with use of an increment borer to extract a pencil-sized core from the tree that exposes the annual growth rings. The increment borer can be found in forestry supply company catalogues. Cores can be glued to a backing block of wood, sanded down, and varnished to expose the growth rings more clearly. The same growth-ring patterns found in increment core samples can also be observed, measured, or counted on cleanly sawn tree stumps (fig. 2.5). Obviously, coring does not require that the tree be cut down. Trees grow better under favorable conditions, usually meaning more available moisture. In drier regions, periods of very favorable growth are interspersed with periods of relatively drier conditions, when fires were more common. These patterns can often be teased from a study of tree rings and give indications of past climatic variation.

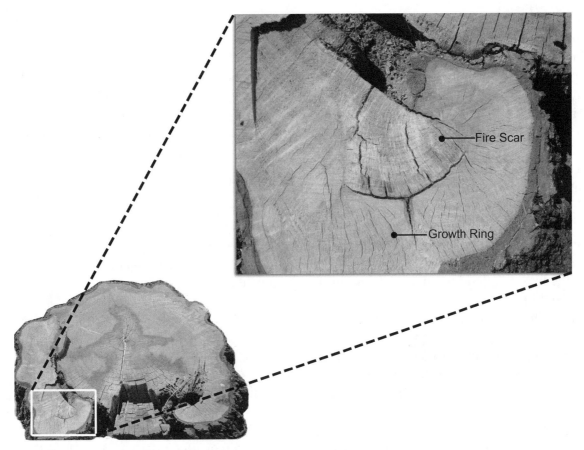

FIGURE 2.5. Cross section of a bur oak stump showing growth rings and a fire scar

Task 10. Map Soils and Surface Geology

Soils integrate the changes that occur in vegetation, soil fauna, climate, and erosion over hundreds of years. Of course, erosion and deposition can occur over hours, but will be recorded in the soil profile for centuries. Soil maps and descriptions of soil types provide the information for beginning this investigation. Compare what you see on the land with the information on the same soil type as mapped and described in the soil survey for your county. (County soil surveys, complete with maps and descriptions of soils, are available

through your county extension office or your county conservationist in the Natural Resources Conservation Service [NRCS] office. Copies also are likely available in university libraries if you have one nearby.) Determine if the soil type is intact or eroded, or if it's been altered in other ways. For example, look for a "plow layer," a more or less homogenous layer at the surface, four to six inches deep, more or less lighter in color than the topsoil is described. This layer is an artifact of plowing, where some of the subsoil was mixed with the topsoil through tillage. If the soil is eroded, expect to find less topsoil or a thinner plow layer. If it has received deposition from erosion occurring upslope or upstream, expect to find the profile buried some depth below the surface. Especially in forested areas, see if you can find charcoal indicating past fire. By comparing soils from tilled fields to those that were protected in fence lines, you can gain insight into the range of variation and how years of tillage or grazing may have altered the soil. See if you can detect differences in compaction or density, for example. When soils have been badly degraded, soil improvement may be a requirement in your restoration before introducing expensive native seeds or plants.

Because soils provide a long-term record of the land history, you can literally dig as deep as you like to learn more about the history of your land. For example, many plants create stone cells, called "phytoliths," and specialized laboratories can analyze soils for the types and quantities. This information could indicate if a warm-season grass, forest, or cool-season grass prevailed on the land. Of course, soil morphology, the organization and characterization of soil layers, provides a good indication of this as well. For example, native grasslands in temperate environments produce thick top-soil with a higher organic content than forests. Deciduous forests lead to thicker topsoil than coniferous forests. Original survey data generally confirm the kind of presettlement vegetation present.

If your interest in your land runs deeper than merely restoration, there are other sources you may want to explore. We only offer this information because using some of these other sources of information can lead to new understandings about your land. We therefore leave it up to individuals on the choice or direction on seeking additional information.

Archeological records, for example, can give you greater background about the history of your land. There also are specialists who study fossilized pollen, primarily in wetlands where these materials are well preserved. This involves a process of collecting samples in relatively undisturbed settings, and after elaborate preparation, microscopically identifying the relative quantities and species of pollen present. These findings can be linked with regional data sets to indicate the types of vegetation that occurred on or near your land long before settlement. In the western states, packrat middens are often analyzed to reveal what grew on the land hundreds or even several thousands of years previously.

Stream-terrace aging can be useful for developing the history of a watershed and stream or river environment. By aging the trees growing on stream terraces, or dating the cultural debris (e.g., soda bottles or cans, types of barbed wire, etc.) found within the stream deposits comprising the terraces, you can gain some insight into the dynamics of the watershed and stream. This same information can be aligned and cross-referenced with maps of surface geology, glacial moraines, and other types of information often available in regional studies. These understandings may be useful to understand the history of a watershed and stream, often helpful in restoration planning around river environments. However, as stated earlier, for most settings, this kind of information may not be particularly useful, at least at this stage of restoration planning.

Putting It All Together

You will now use the information you have gathered to interpret the past. The purpose is to learn what has happened to your land and use that knowledge to inform your restoration plan. Understanding the change agents—those conditions and actors responsible for the shifting ecological conditions and ecosystem health, from historic to current conditions, helps crystallize the relationships and strategies that may be desirable and necessary to effectively restore your land. This understanding is best derived from the investigations described earlier.

Task 11. Understand How Your Land Has Changed

Because the available tools for evaluating history, such as aerial photographs, do not cover every year, or provide insight to every aspect of your project, there will be gaps of understanding and leaps of faith required to pull puzzle pieces together into a restoration plan. As you review what you have learned, think at both the scale of your project and the scale of the broader neighborhood. For example, perhaps a forest was cleared for agriculture on your land, while on a broader scale agricultural development began to reduce the frequency and intensity of historic wildfire. Both have some bearing on your decisions regarding a restoration plan. Thinking at the two scales also helps you to recognize reference areas where historic conditions and relationships may be still present and observable on your land or on others. The identification and use of reference areas can be very useful for learning about the historic conditions on your land. During Steve's synthesis of available information for his farm, he found not only a reference area across the street, but learned of other such areas, often in dedi-

cated nature preserves within his neighborhood. These all were visited and became a powerful source of knowledge for restoration of his own farm.

This change analysis is completed by overlaying the tracing paper map of existing and summarized historic conditions to create a single additional tracing paper record that shows the details of where and when changes have occurred. Figure 1.6 shows this change analysis conducted for Stone Prairie Farm, by Steve. He followed the same tasks in steps 1 and 2, and used historic analyses to create this synthesis and summary in that figure.

Historic investigation and change analysis are usually very revealing. These processes are ongoing and will continue to open new insights. For example, only this winter, twenty-five years after starting restoration on his farm, Steve learned from the son of a neighbor that prairie chickens used to roost in willows along his creek. The largest bur oak in the neighborhood, some 60 inches in diameter, awaits tree-ring analysis.

Change analysis readies you for developing working hypotheses on how the ecosystem previously functioned. Within the context of existing stressors and controlling influences, how has the land changed? The more you understand, the better positioned you will be to develop a good restoration plan.

Notes

1. May Thielgaard Watts, *Reading the Landscape: An Adventure in Ecology*. (New York: Macmillan, 1957). Reprinted by Nature Study Guide Publishers, Rochester, 1999.

Interpret Landscape Changes

It takes as much energy to wish as it does to plan.

Eleanor Roosevelt

In this step we walk through the process of summarizing what you have learned from previous investigations of your land, leading to questions and assumptions about how the ecosystems once functioned and how they have changed. Because you are basing your assumptions on interpretation of incomplete historic data, there will be much uncertainty. Everything we think we know is based on assumptions, some with a good basis in facts, but most less so. The more thoroughly you can explore historic information and the more completely you can fill in missing pieces, the more accurate your assumptions will be. These assumptions and the associated questions will be worded as working hypotheses about the ecosystems in your project, how they looked, how they functioned, key species and their distributions and relationships, and so on.

Through activities in previous tasks, you have completed the following, as far as possible:

- Investigation and mapping of current and historic conditions on the land
- Changes in the land and reasons they occurred or continue
- Assessment of past and present stressors responsible for the changes

Working Hypotheses

Before we begin preparing a restoration plan, we need to develop working hypotheses and principles for the restoration. The assumptions we make about the historic condition and functions of the ecosystems in our project also are key to identifying the endpoints or goals; what do we want our restoration to achieve on the land. This latter point will be addressed later. First, we will focus on the assumptions and hypotheses.

Task 12. Develop Working Hypotheses

You can never be certain what ecosystems existed on your land before it became altered by recent humans, or exactly how they functioned. Yet, this prior condition is the basis for developing good restoration plans. Using the insights you have developed, you can begin to make assumptions about prior conditions and how they have changed. These assumptions are working hypotheses, an important step in restoration planning, not just at the beginning, but throughout the years of the project. If your underlying assumptions are not clearly articulated, your goals and objectives may be off target and lack a valid context. Also, developing your hypotheses with other stakeholders and sharing them widely with everyone involved in the project helps to keep the project focused.

Historic ecosystems were acted upon by stressors that lead to existing conditions. In addition to the historic ecosystems and how they functioned, it also is important to include hypotheses about how stressors historically altered the ecosystems as well as how stressors continue to affect the ecosystems of your project. The transition from historic conditions is typically driven by many stressors, as well as a sequence of stressors (e.g., initial clearing of forests and livestock grazing, followed by changes in hydrology, tillage, and eventually by chemical weed control). These changes occurred in different locations often in very similar sequences and patterns. You previously assessed historic conditions but now need to consider current stressors. This is not as difficult as you might think. There are very few types of stressors (see box 3.1) although endless degrees and variations of them. We will review some basic ecosystem principles that will help you understand the ecosystem functions that may be affected by the stressors. The effects of stressors are manifested through ecosystem functions, and understanding the functions allows you to better evaluate the stressors.

Ecosystem Characteristics and Principles

Nature is organized and operates in some fairly consistent ways. Certain characteristics are found in nearly every ecosystem but often are obscured or confused by the effects of the stressors or changes resulting from agriculture and development.

Ecotones and *gradients* are the transitions between different ecological communities or ecosystems. They may be an abrupt demarcation, such as where a steep upland soil type meets a poorly drained wetland margin. In such a setting, the transition between the two ecosystems may occur within a distance of a few feet to even a few inches. In contrast, on slopes above the wetland, the transition may be very gradual, often stretched out over hundreds of feet, making it rather arbitrary as to where one community ends and another begins. These transitions are called ecotones, and they reflect the steepness of environmental gradients.

The types of gradients that contribute to ecotones include slopes, aspect (direction), parent soil types (e.g., type of rock or unconsolidated material such as gravel, sand, or silt), hydrology, disturbance regimes, and microclimates. These are not independent. For example, aspect is an important determinant of microclimate but also disturbance regime. Fire was more frequent and more intense on south-facing slopes in the northern hemisphere, for example. Depressions and draws receive cold air from ridges and slopes, and this is commonly reflected in vegetation. The cold species will occur more commonly in depressions where cold air settles. Snow accumulation and melt ecotones are widely present not just in the mountains where snowfields recede as spring advances, but also in grasslands of the Great Plains where snow accumulates in draws and drainageways. Shorelines represent yet another common gradient (figure 3.1). Ecotones resulting from erosion and sedimentation also are common. Scree and avalanche slopes are often present in mountainous areas. Rivers and associated floodplains also have patterns of vegetation resulting from flooding, sediment deposition, and scouring.

BOX 3.1. BASIC CATEGORIES OF STRESSORS WITH EXAMPLES

These examples are not independent; change in one often results in modification in others.

- Altered hydrology
  1. Increased runoff because of disturbance in natural vegetation.
  2. Ditching.
  3. Tiles to dewater poorly drained soils.
  4. Deepening and straightening stream channels.
  5. Altered flood regimes because of dams, culverts, etc.
- Physical stressors
  1. Tillage of various types.
  2. Loss of topsoil, usually through erosion.
  3. Restructuring of slopes with terraces, diversion ditches, or alteration by earth moving.
  4. Compaction, usually from intensive grazing and/or tillage.
  5. Decreased oxygen levels in streams and ponds, usually from eutrophication.
- Biological stressors
  1. Invasive species.
  2. Loss of native species, including soil microflora.
  3. Dilution of local genotypes.
  4. Nonnative parasites and/or diseases.
- Chemical stressors
  1. Fertilizers, both in soil where applied, and in runoff.
  2. Change in pH, usually from application of lime or change in hydrology.
  3. Depletion of trace minerals.
  4. Nitrogen enrichment of soils and water.

Task 13. Map Ecotones and Gradients

Based on what you have learned about your land, map what you believe are the key gradients that historically occurred. Refer to the following gradient categories to create the gradient analysis data form (data form 3.1, appendix 1), and on your basemaps define the areas where you believe important gradients and associated ecotones occur. For each, draw arrows or vectors showing the direction of the gradient. For example, for a slope gradient the vector arrow might start at the top of a ridge and run down the slope to the bottom. Then attempt to create a rough map of the ecotones, separating vegetation types that you believe occurred along each gradient. This has to be crafted from what you learned about the historic conditions present on your land, or the reference areas you visited in your neighborhood. Map those and others that you might know about

FIGURE 3.1. Gradients in water depth and plant community types such as along ponds, lakes, and rivers/shore-lines are nearly universal

or become aware of during your local investigations of the history and visits to reference natural areas.

- *Slope change or break ecotones*—the vegetation growing on steep slopes and level benches will vary considerably.
- *Soil type ecotones*—high and dry sandy soils will have drought-adapted plant species, while clay soils may have species best suited to cooler, damp conditions.
- *Parent material ecotones*—plant species types may vary depending on the type of geological substrata present. Clear examples occur where bedrock depth varies from near the surface to deeper depths, or soils that develop over sandstone versus limestone or shale.
- *Slope aspect ecotones*—south/west slopes and north and east slopes typically support plants found in warmer/dryer to cooler/wetter settings, respectively.
- *Hydrologic ecotone*—pattern of different types of plants associated with water availability, from permanently saturated conditions to excessively drained conditions.
- *Snow-melt pattern ecotones*—concentric rings or parallel bands of vegetation growth and maturation reflecting average time melting exposes the soil.
- *Fire-gradient ecotones*—zones from areas that burn more frequently and hotter to areas that burn less frequently and less hot, often associated with aspect.
- *Frost pockets*—low areas, or openings in forest where reradiation or cold air drainage results in cooler conditions at night, and more frequent frost.

• *River and floodplain ecotones*—areas that are flooded, with progressively less flood impact as elevation increases away from the stream.

Food webs are a characteristic of all ecosystems—who eats whom. From your hypotheses about the major species that previously existed in your ecosystems, you can begin to develop food web diagrams that provide additional insights into how those ecosystems functioned. Figure 3.2 summarizes presumed historic relationships in the tallgrass prairie ecosystem, and how they have changed. Sketching food webs is a good example of how one set of hypotheses leads to additional insight and further hypotheses. When mapping the food web, place what you believe to be the most influential group in the center of the web and then create a primary and secondary ring of groups, with each succeeding ring of lesser importance, as shown in figure 3.2. The illustrated web map was created to illustrate assumptions about the major changes in the prairie

ecosystem as the prairie was fragmented by settlement and farming. The assumptions of which groups of organisms were present and their relative importance are based on research similar to what you have completed for your project. In reality, food web relationships, including relative importance and position of groups, may change in different geographic settings.

If you want to become really absorbed in unraveling the food web, you can develop species-level mapping of relationships. This requires considerably more research and knowledge than using groupings of species. Species-level mapping would start with listing each species of animal and plant present and mapping their relationships and relative importance, linking each to the others depending on relationships such as predator-prey. Pollinators can also be important to consider. Other relationships include nonfood habitat requirements, such as where birds nest and where hibernating organisms overwinter. There is almost no end to the ways in which you can begin to think about how species relate to

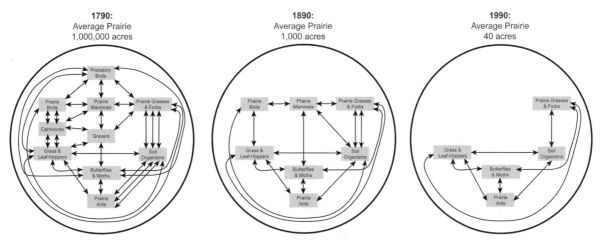

FIGURE 3.2. Changes in the prairie ecosystem presettlement to present, illustrated with food webs

one another. The more you investigate, the better your insight into how the ecosystem worked that you are planning to restore.

Specific species-level mapping can be especially important for certain groups, such as butterflies and pollinators. They commonly have specific host species for their leaf-eating larvae and flowers, often of different plants, for adult foraging. Successful restoration of their habitat cannot be done without both groups being restored.

Optional Task. Sketch Food Webs for the Ecosystems in Your Project

Succession is yet another characteristic of all ecosystems. Nature is dynamic. From day to night, season to season, year to year, with climatic variation at scales ranging from decades to millennia, patterns of life are never static. The patterns pass across landscapes with both subtle and profound variation, but usually change slowly relative to human attention or life-spans. These patterns, their temporal and spatial scales, and how they occurred on your landscape provide important insight into the functions within your ecosystems. Indeed, much of the change is driven by stressors, so understanding the changes is frequently a key to getting a firm grasp on the stressors.

Two important and fairly obvious drivers of change are flooding and wildfire. Both are tied to climatic variation and landscape position, and dynamically affect plants and animals. Flooding, of course, will primarily affect ecosystems bordering streams or depressions, while fire is more important on uplands. However, fire can also be very important in maintaining wetlands. Response to these, or any disturbance, is called suc-

cession. Changes in the biological, physical, and chemical environments of an ecosystem resulting from the disturbance, shifts the balance to favor some species and disfavor others, resulting in changes in relative numbers or even presence and absence of species. Over time, these changes trend toward restoration of the predisturbance conditions, and communities very similar to what occurred will return, if the ecosystem is reasonably healthy. A different trajectory of succession may occur, however, if change in hydrology or extirpation of keystone species has occurred. If that is the case, an alternative trajectory is set into motion, leading to a different endpoint. Depending on the extent of the disturbance, the response cycle can require years to centuries. One way of thinking about ecological restoration is to make sure it is on the right trajectory, and jump-start the succession, reducing the time that the ecosystem would require to repair itself. Recognizing the successional pattern and how to manage and direct the trajectory will guide your restoration. Restoration treatments are designed to trigger successional responses.

Historic patterns on a landscape, especially where the land is greatly altered, particularly by agriculture or development, are not easily discernable. To gain a clearer picture of where succession would go can often be determined only by examining comparable natural areas. This is one of several reasons why reference natural areas are extremely valuable to your planning. The effects of agricultural conversion on ecosystem patterns are clear in this side-by-side image of floodplain land along the Sugar River, southern Wisconsin (see fig. 3.3).

Each species needs critical space. All species have fairly clear relationships to the size of the habitat they require. This is easy to understand

FIGURE 3.3. Comparison of altered and intact floodplain land, Sugar River, Albany, WI

for a grazing animal such as a bison or a cow. Ranchers are keenly aware of the acreage needed to grow the required quantity and quality of food for a cow. In some semiarid regions, a cow may require fifty acres of range to remain nourished. In more humid regions, two to three acres of pasture, or less, may be sufficient.

It is harder to understand what area predators need, but they also have relationships with land area. Their relationship, however, is less direct and depends on populations of prey. The prey species, in turn, have area needs depending on their primary food species, mostly plants. Pack animals like wolves have area needs related to their social structure, and their range must sustain the entire pack.

Figure 3.4 summarizes some of the documented land area needs for some groups of wildlife. You might consider this land area pyramid as a working hypothesis. The area needed by any species changes from year to year, and region by region depending on weather and changes in food availability, but the generality holds. For example, if your restoration project is five acres, a goal for your restoration plan cannot be restoration of wolf habitat. Instead, you need to focus on habitats that will support native ants, butterflies, birds, amphibians, or small mammals.

You now should complete task 12. Do not be concerned that you have many unanswered questions for which you lack even good guesses. It is more important at this stage that you state your assumptions. What did this ecosystem look like before white settlers began clearing for agriculture? What were the dominant or keystone species? How often did it burn or flood? For as long as you continue to observe and work with your land, you will continue to ask more questions, but also, you will get more answers. This task, therefore, is open ended, but not one to be skipped as you begin to develop a restoration plan.

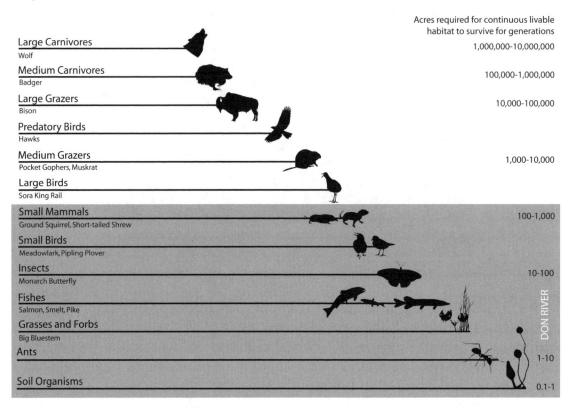

FIGURE 3.4. Pyramid of life in the tallgrass ecosystem

Summary

Everything we know, or think we know, is based on assumptions. In ecological planning, it is important to recognize our assumptions. These can be stated as working hypotheses, and through your investigations you can often verify, or at least firm up, your assumptions. With step 3, we try to state as many of the assumptions, or hypotheses, as possible about the ecosystems in our project. Some assumptions or hypotheses will be tested through monitoring of results or experimentation designed into your restoration plan (step 6).

Step 4.
_____

Develop Goals and Objectives

Without goals, and plans to reach them, you are like a ship that has set sail without a destination.

Fitzhugh Dodson

To this point, we have focused on learning about the land. By following the process and tasks in the previous steps, you should have some detailed understanding of the land including the ecological health of the component ecosystems. What and where are the stressors, how have they acted to change the ecological health of the ecosystems, and what kind of restoration work is needed to move these systems back to a healthier condition? The depth of your understanding of the land should have been increased through review of reference areas, larger landscape-scale historic events, and refined through consideration of the agents of change.

Building on what you have learned about your property from earlier tasks, you now need to set goals and objectives, and begin planning. In this step, we cover the following components of planning:

Goals and *objectives* indicate your intentions and expectations, including endpoints of the remedial restoration program, conditions that can be confirmed by monitoring.

Framing steps include leadership, governance, and decision processes. These are critical if you are doing the planning with a group, whether your family, a volunteer organization, or employees.

Preliminary budgeting is necessary because budgets drive planning in more instances than not, and your goals and objectives need to be within your means.

Task 14. Develop Restoration Goals and Objectives

The goals and measurable objectives for a restoration plan initially should focus on the *stressors*, both those still affecting the ecosystems and those that in the past led to a loss of ecological integrity. Stressors will likely vary among the different ecological units or cover types you have identified in your project, especially if your project is large or complex. Thus, you may have as many sets of goals and objectives as you have ecosystems in the landscape.

A simple chart (see data form 4.1, appendix 1) can help you align goals and objectives with different ecosystems. This is particularly useful for discussions with others involved in the project.

53

You will link stressors spatially to the maps you have completed in steps 2 and 3. The stressors and locations also should be tied to specific tasks you think necessary to achieve the measurable objectives. In a later step, we will add additional columns to the table to assign schedules, responsible parties, costs, and monitoring activities to be able to evaluate progress toward objectives.

While the formulation of goals and objectives may be difficult initially, it should not be a stumbling point in moving forward with a restoration plan. The worksheet will help organize and structure the connections between goals, objectives, and the tasks to achieve them. Keep in mind that goals and objectives are not inflexible. Monitoring (step 6) will be an ongoing process, and it includes a systematic review of goals and objectives, with adjustments as you learn from both successes and failures.

Sometimes, especially with larger groups, brainstorming can be helpful as a way to start identifying goals and objectives. For example, box 4.1 shows the result for land managers who brainstormed the general ecological objective statements for several properties they managed. They organized them using the tabular format suggested earlier.

Another example was a family group where the father was the only one really focused and somewhat knowledgeable in ecological restoration. He led his family and some neighbors through a focused exercise that began with listing problems. The problems, of course, were later defined as stressors and led to development of *guiding principles* (goals). During the process, everyone came to a common understanding of how their land-use activities contributed to the stressors on adjacent properties, inevitable because all neighboring properties were hydrologi-

---

> **BOX 4.1. NATURAL AREAS MANAGEMENT PROGRAM OBJECTIVES**
>
> - Stimulate existing native seed bank
> - Stimulate native ground cover
> - Reduce nonnative vegetation
> - Reduce soil erosion and sedimentation
> - Promote gradients of size and age structure
> - Promote vertical structure gradients
> - Restore surface and subsurface hydrology
> - Restore population dynamics
> - Restore continuity between systems
> - Create opportunities for human use and appreciation
> - Create educational programs to increase awareness
> - Create participatory programs to build proprietary interest
> - Create opportunities for dispersal of species
> - Reduce fragmentation

---

cally connected (see fig. 4.1). The family farm that initiated the restoration was at the receiving end of eroded soils, nutrients, and manure management problems largely from neighboring properties. What could have been quite contentious turned out to be a matter-of-fact process that was very productive. The neighbors heard the goals and objectives for the family farm, and realized their partnership was not only needed, but that they would also benefit from successful restoration of the neighborhood.

GUIDING PRINCIPLES - LEOPOLDIAN CONCEPTS/HYPOTHESIS

1. Improve land health
   a. "The capacity for self-renewal in the biota"
2. Enhance ecological integrity
   a. "The presence of a full complement of native components in their characteristic numbers"

OBJECTIVES

1. Protect, maintain, enhance biota
2. No bluff land area development
3. Possible development in other areas (nursery building, etc.)
4. Restore productive apple orchard with possible heirloom varieties
5. Maintain private ownership
6. Preserve farm and family history, gravesites, Indian mounds, etc.
7. Improve fishing quality of pond
8. Provide educational opportunities;
   a. By example
   b. On-site place to experience, see and learn about restored systems
   c. Create strategies to tell the story of the experience
9. Develop plans and program that monitor success and performance

FIGURE 4.1. Goals and objectives created for restoration of forests, stream margins, savannas, and prairie lands at a private ranch, IA

Externalities That Potentially Impact Goals and Objectives

Goals and objectives must be adjusted for externalities. There are many external factors to be considered:

- regulatory or permit issues and requirements that may take time before a particular task can be implemented (e.g., restoring a stream or wetland that has been partially channelized or filled);
- lack of agreement among neighbors or family members either with the overall objective or some parts;
- time required to build partnerships with neighbors and others whose involvement is necessary, or whose involvement improves your project;
- permissions required from agencies or stakeholders who will be affected;
- costs and budgetary limitations;
- available time (your own or that of others) to effectively implement and follow through with the project;
- availability of seeds and plants of appropriate provenance, genetic sources;
- existing contracts or leases;
- existing timber management contracts.

Make sure you absolutely understand the laws and regulations that might affect your land and your plans. Your county conservationist or NRCS office can usually advise you. In most locations in the United States, any direct or indirect modification of water-drainage features, septic or sewers, wetlands or floodplains, floodways, work on steep slopes, timbering, plowing of ground not plowed (at least since 1985 or before), any change or diversion of water uses and rights, should all trigger a potential regulatory flag for your project. There are other issues as well, such as shared tile lines or drainage ditches, even fence line changes between neighboring farms. Regulatory issues are an area where you may need to consult one or more authorities with experience.

Practical considerations. When generating goals and objectives, it is common for people to corrupt the process by practical considerations such as available funding or labor. This results in an incomplete or diluted set of goals and objectives that fail to address all the stressors. In step 1, we discussed how externalities of any type (e.g., exploitive resource uses, financial motivations, recreational desires) can significantly affect the ability to successfully restore ecological health to the land. We encourage you to follow the process laid out above and generate a comparable table as in data form 4.1, appendix 1. This is not to be delusional about externalities, but simply to ensure that you see the whole restoration picture before you create or revise a restoration plan. In the following sections we account for externalities, such as available funding and labor.

Addressing Uncertainty

When developing a restoration plan, you likely will be uncertain about how to address one or more specific problems or stressors. You should have listed all stressors in the table, and developed what we will call *preliminary goals, objec-tives, and tasks*, but you may be uncertain about what will work. If so, make a note of your uncertainty so that you do not lose track of it. Perhaps consultation with someone with experience is needed, or maybe you will have to investigate potential solutions more fully, or approach neighbors to help address an off-site problem. When you are unsure about whether a particular treatment will work as you hope, we suggest test plots or trails on a small scale before committing more time and money. Regardless, do not dismiss the problem just because you see no solution at this point.

When there is a high degree of uncertainty, you should not commit to a restoration action or task. That is especially the case where there is disagreement within the group about the problem or proposed solutions. When this occurs, we have found it useful to continue the process by taking a different tack for the areas surrounding the uncertainty, indecision, or disagreement. First, define what is uncertain. Why is there indecision or disagreement? Attempt to gain as much clarity and consensus among all parties as possible. Second, decide if the uncertainty arises from fundamental disagreement, or simply a lack of experience or knowledge, which is commonly the cause of disagreement. If the latter, then seek more information. If the former, then explore the underlying reasons for the different opinions, and examine the assumptions in each case.

There will always be some uncertainty, and this is a good time to reflect on how much is acceptable with your project. The level of uncertainty that you are willing to accept usually involves money, time, and labor. If the uncertainty is too high, look for ways to reduce the risk. That usually involves getting more information and using that information to develop alternative strategies. For example, if you are planting native grassland, there is a risk that a droughty spring could lead to failure. This is not unlike what farmers experience every year in planting their crops. If there is a higher risk that the plantings might fail because you missed the earliest possible spring planting window that normally would allow seeds to germinate and establish before the drought, the feeling of uncertainty is appropriate. You probably should reduce the risk by either not planting until fall or plant half the area now and half in the fall, thereby spreading the risk and gaining more information about the best approach. Perhaps you can substitute lower-cost species for the spring planting, and plant the more expensive species in the fall. In other words, once you identify the uncertainty and investigate the cause, solutions may become more apparent.

Test and demonstration plots. You obviously can reduce risks, such as planting at the proper time, but only if you know enough to recognize potential solutions for reducing risk. Test plots are an inexpensive way to reduce risks. Uncertainty arises from lack of knowledge or experience, or disagreement about the best way to do something. The goal is to understand how the system responds to different treatments, or how much time, effort, and costs are involved. The better you have defined the specific points of uncertainty, disagreement, or risk, the easier it will be to design and install useful test plots.

For each area of uncertainty, it is useful to come to consensus on the best- and worst-case outcomes. For example, if someone on the restoration team is opposed to using herbicide to treat an invasive shrub, you might consider a test plot that involves doing nothing, and compare it with manual cutting, and cutting plus herbicide. It is nearly always useful to carefully consider the assumptions for worst, best, and intermediate scenarios to evaluate the understanding that underlies different opinions (see fig. 4.2).

Determining the effectiveness of management treatments is the most common use of test plots. You should also consider plots to evaluate the costs and labor requirements before implementing restoration at larger scales. The size of test and demonstration plots should be appropriate for the scale of the problem you are investigating. For example, if you want to test effectiveness of different herbicides, a few square meters might do, but if you want to know what time of year is most effective for a prescribed fire, a few acres would be more appropriate. (Data collection techniques described in step 6 should be used to compare results. Pretreatment baseline data compared to posttreatment results are better than only posttreatment data.) Figure 4.3 provides several test and demonstration plot designs that we have used for evaluating restoration effectiveness before scaling up. The figure shows test plots with dimensions, goals, and sampling design for the baseline and follow-up sampling. Specific methods were established to sample key variables at each sampling time period. The data were summarized and used by the restoration team to make informed decisions on technologies, restoration process, costs, and labor needs for implementing restoration. These plots were also effective for community education and consensus

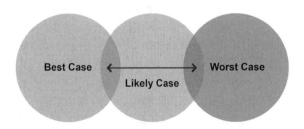

FIGURE 4.2. Framing the treatments for test and demonstration plots

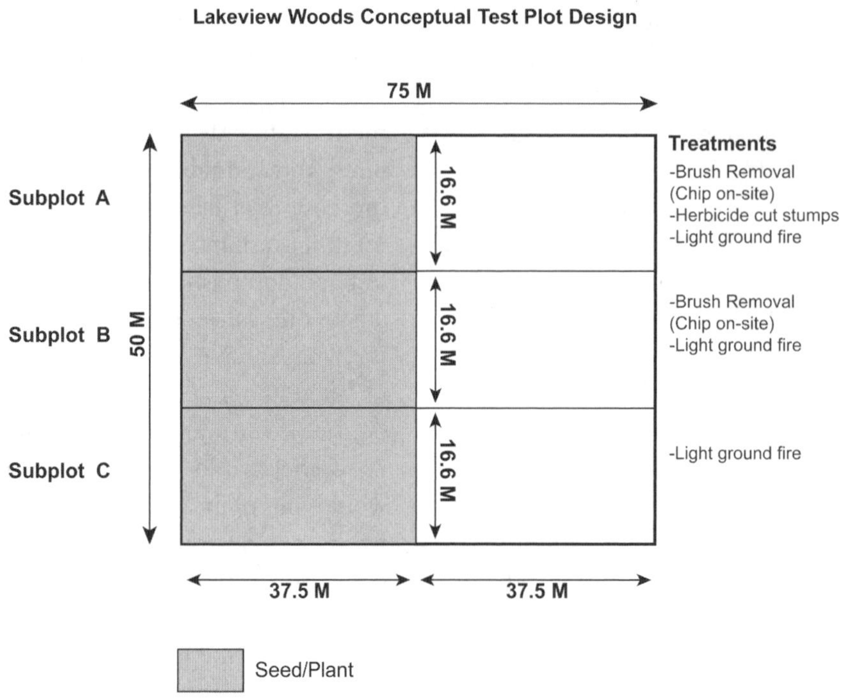

Lakeview Woods Conceptual Test Plot Design

Treatments

Subplot A
-Brush Removal
(Chip on-site)
-Herbicide cut stumps
-Light ground fire

Subplot B
-Brush Removal
(Chip on-site)
-Light ground fire

Subplot C
-Light ground fire

75 M

50 M

16.6 M

37.5 M   37.5 M

Seed/Plant

Treatment Schedule	Year 1				Year 2				Year 3			
	1	2	3	4	1	2	3	4	1	2	3	4
Brushing				X				X				
Herbiciding				X				X				
Burning						X		X	X			
Monitoring											X	
Reporting				X				X				X
Seeding/Planting						X	X	X		X	X	X

FIGURE 4.3. Test plot layout at a Wisconsin woodlot restoration project

building, as results were clearly visible. However, some treatment effects are only statistically apparent and require analysis of baseline, control, and posttreatment plots. One should not assume that results that can be seen demonstrated through data analysis are unimportant; some subtle differences after a year or two can become extremely important in five or ten years.

Any time you use test plots, make sure you allow adequate time for the treatment effects to develop. Responses sometimes take a few years. Be

both realistic and practical about timelines. We typically run test plots two to three growing seasons and longer when we see subtle differences developing.

On a savanna restoration in Wisconsin, there was serious disagreement over using herbicides to treat invasive shrub and tree stumps. Some had read that coffee cans placed over cut shrub stumps provided a nontoxic alternative. A restoration technology decision matrix (table 4.1) was designed to enable all parties to examine

TABLE 4.1

Restoration technology decision matrix

	Labor[a]	Cost/hr[b] $10, $50, $500/hr	Treatment acreage	Estimated cost (relative)	Work capacity, % volunteer services[c]	Probability of success w/volunteers[d]	Total score	Relativized score	Human risk factor[e]	Human exposure factor[f]	Adjusted score
WOODY					Annual						
Girdle	100	50	8	$40,000	0.1	1	4000	1.15	2	1	2.3
Girdle and herbicide	110	50	8	$44,000	0.1	1	4400	1.27	2	1	2.5
Cut, chemically stump treat, let cut stem lay	55	50	8	$22,000	0.1	1	2200	0.63	2	1	1.3
Cut, chemically stump treat, drag and chip	100	150	8	$120,000	0.1	1	12000	3.46	2	1	6.9
	500	10	8	$40,000	1	3	120000	34.63	3	2	207.8
Heavy wood chip mulch	100	50	1	$5,000	1	3	15000	4.33	2	2	17.3
Landscape cloth mulch	40	500	1	$20,000	1	3	60000	17.32	1	1	17.3
Repeat cutting	165	50	5	$41,250	1	2	82500	23.81	2	2	95.2
Chemical wick application	10	50	8	$4,000	0.1	1	400	0.12	1	1	0.1
Goat herbivory	200	50	5	$50,000	0.1	3	15000	4.33	3	3	39.0
HERBACEOUS				$0							
Pull	50	10	5	$2,500	1	2	5000	1.44	1	1	1.4
Mow	4	10	2	$80	1	1	80	0.02	1	1	0.0
Chemical wick application	10	50	2	$1,000	0.1	1	100	0.03	1	1	0.0
Competition plantings	10	10	2	$200	1	1	200	0.06	1	1	0.1

TABLE 4.1

Continued

	Labor[a]	Cost/hr[b]	Treatment acreage	Estimated cost (relative)	Work capacity, % volunteer services[c]	Probability of success w/volunteers[d]	Total score	Relativized score	Human risk factor[e]	Human exposure factor[f]	Adjusted score
MIXED				$0							
Biological control	5	500	10	$25,000	0.1	2	5000	1.44	2	2	5.8
Virus infection	5	500	5	$12,500	0.1	2	2500	0.72	2	2	2.9
Goat herbivory	200	50	5	$50,000	0.1	3	15000	4.33	3	3	39.0
Prescribed burning	4	200	8	$6,400	0.1	1	640	0.18	2	2	0.7
Bacterial or fungal infection	5	500	5	$12,500	0.1	2	2500	0.72	2	2	2.9
TOTALS							346520	100.0			

Note: Highest score = highest labor, highest cost, highest human risk, and highest human exposure

[a] Estimated person hours per acre is based on ~ 2,000 stems/acre requiring treatment

[b] Time for initial and follow-up treatments for 2,000 woody stems/acre

[c] 1=100% vs 0.1=10 %; 1=90% professional. "1" means 100% of the work is to be done by volunteers.

[d] l=3, m=2, h=1

[e] l=1, m=2, h=3: Human risk: injury from use of tools (hammers, nails, saws, etc.) that could result in injury or fatality

[f] l=1, m=2, h=3 Human exposure: short- or long-term injury from use of chemical controls (e.g., herbicides, road salt, aerosol paints, etc.) and biological agents (e.g., biological control bacteria, fungi and viruses, etc.). Where trained personnel in proper safety equipment using herbicides according to the MSH procedures are doing the application of chemical controls, human exposure and risk as defined by USEPA, USDA, and others is defined as low to very low to not measurable. Exposure risk to users of parks is considered here also as follows: where discriminate use in very small quantities are directly applied once per season to specific plant stems that are removed (> 10 meters) from public-use trails, then exposure risk is for purposes of this exercise assumed to be very low to not measurable. Risk is only higher than very low when unlicensed users start doing the application or have access and a very high probability of contact with treated plant tissue within 24 to 36 hours of application. Contact is defined as direct skin contact with virulent versions of the herbicides in sufficient quantities to cause or contribute to known, documented health risks.

60

economics, labor, restoration efficacy, and minimization of exposure to herbicides for the alternatives being considered. In this example, the numbers are estimates, to provide a relatively unbiased comparison of costs and benefits. Decision matrices can be refined by incorporating data, such as real costs or time requirements to apply each method from test plots. In fact, this type of matrix is often used with test plot data to compare alternative approaches (see step 6). The savanna group was able to reach agreement using the matrix, which confirmed that the can method would be too labor-intensive, primarily because it required multiple trips to cut sprouts that escaped the cans. That was confirmed with a small plot study.

Good Framing Infrastructure

We have watched from a distance as restoration projects were abandoned and asked ourselves why, with such a good enthusiastic start, did the project swerve off track so quickly? When asked to help pick up the pieces, we learned that projects commonly failed because of those who were leading them:

- They were inspired about working outdoors with nature, but had unrealistic notions about the effort restoration requires.
- They did not take the time to communicate and share aspirations, or to develop the platform for sustaining a long-term project on the ground.
- They made incorrect assumptions about the future, about who would caretake or maintain a project.
- They failed to formalize a decision-making process that was durable.
- They were replaced without proper orien-

tation of the new leader, or new leaders came with strong opinions for changes that sometimes included underlying philosophy, technical solutions, and/or priorities.

Task 15. Review and Firm Up Framing Infrastructure

Here we focus on ensuring that you have the framing infrastructure well established, with broad understanding among stakeholders, and a clear process for making changes in the restoration program. Whether you are working on public land or private land, with your family, volunteers, or paid employees, these are critical foundation stones to help people work more effectively in restoring the land. The intent is to lay a strong foundation of understanding to guide everyone involved with the project, including some who may come into the project in future years. During this task you should revise your thinking and commitments to your restoration program accordingly by thinking about the common causes listed earlier that often result in the failure of restorations. As you think through your specific project, record the uncertainty you have about how your plan will address each of these common causes of failure. This involves development of what we call a *governance plan* for your project, defined in the next task. Lack of governance planning is often the single most common reason restorations become side tracked or simply wither and die.

Task 16. Develop a Project Governance Plan

The governance plan becomes a critical part of the final restoration plan. Restoration is most successful when governance is included from the outset, throughout the entire design process. The following considerations need careful attention at this time (as well as in later tasks):

1. *Affirm the decision to move forward with restoration planning.* In this step, initiators and stakeholders affirm their commitment to explore the restoration. This results from conversations through which all participants acknowledge the need for restoration and agree to begin developing the plan. This can begin with sharing definitions of restoration, discussing discoveries of stressed or failing ecological health on the land, how the land might be affecting water quality, uses being made of the land now and in the future, and wildlife species and abundance, among other considerations. There also may be economic considerations, such as poor fences or failing tiles with implications for financial investments, or changing needs and uses of the property. The test of success for this affirmation and commitment process is the development of strong alignment between the needs of the land and investors, landowners, volunteers, or others on the team, and also between the needs and commitments required to restore the land. Many restorations fall apart when the time commitments are not understood, or expectations of different parties start out unaligned. Getting this alignment right is an important first step.

2. *Clearly establish responsibilities, and put them in writing.* Once stakeholders have agreed on beginning a restoration and have committed to their responsibilities, it is essential to define how those involved will work together. This involves structuring the leadership and decision-making process so that everyone knows and agrees to their roles and how decisions will be reached. This is especially critical when the proj-ect involves public land or land owned by an organization where leadership is likely to change over time. Having the administrative authority for the land pass a resolution that affirms their commitment, the agreed-to principles for the project, and a schedule of reporting and annual review of the project can save a lot of grief. If there is any concern about not having such commitments over time, it may be worth reconsidering initiation of restoration work on such lands. A lack of commitment resulting in an aborted project could result in more ecological damage to the land. We have seen public agencies agree to a proposed restoration plan to cut nonnative bush in the process of oak savanna restoration by volunteers. The work began when a vocal minority objected to the use of herbicides to prevent resprouting of stumps. Lacking an agreed-upon protocol and leadership authority, herbicide treatment was abandoned, the brush resprouted, and for every cut stem, eight stems soon developed. Savanna restoration not only failed, but the ecosystem was further damaged.

3. *Develop and get agreement on leadership principles and process.* Leadership is critical to the success of any project. Even if working on your own land, it is remarkable how restoration can struggle or fail when other family members have strong opinions and do not accept or support the leader. In a restoration of a northern hardwood forest, for example, the plan called for restoring old-growth characteristics in a secondary forest by thinning to adjust composition and age classes. The project was going well until one son questioned the principle of restoring a forest by selection thinning. He was a deer and ruffed grouse hunter who wanted small clear cuts that would create more young-growth aspen stands favoring those wildlife species. The resulting conflict caused nearly a ten-year delay because the

family lacked leadership to resolve these differences in what would have been one of the more critical growth stages for restoring the forest. Consider that while some members of the family may not initially have an interest, that can change. Also, keep in mind that restorations often take decades, and leadership and membership will change, even in families. Interests in projects can also change as members hear from others about exciting results or concerns.

Strong leaders communicate intentions, engage others, and share information so that everyone is well informed about aims and progress. They regularly review goals and methods with participants to make sure they are still appropriate. If all members of a restoration accept and understand this cycle, they are more likely to (a) participate; (b) follow the direction of the leadership; and (c) passionately or dispassionately appreciate and understand the interim changes and the long-term results; (d) be patient with expectations; and (e) defend the changes in the land during the restoration process, including the often not-so-aesthetically pleasing steps associated with some tasks, such as prescribed burning, logging, and brushing.

4. *Make sure goals and objectives are clearly articulated and agreed to by all participants.* One of the most useful ways to insure good governance is to keep the focus on the ecological health of the land. Show and discuss examples of healthy areas and areas that are not healthy. In *Restoring Ecological Health to Your Land* we discussed the indicators of ecosystem health. They may be useful to help define your project. Work to achieve consensus around a simple philosophical approach, and for the underlying principles and goals. In the example of the grouse hunter and his family, the philosophical aim of moving the forest ecosystem toward old-growth characteristics was rejected by

one member of the family, resulting in delay and lost opportunity. Figure 4.1 illustrates not only the goals and objectives, but also, the philosophical approach adopted by a family. The family started the process by asking all members to read Aldo Leopold's *Sand County Almanac*, which was then discussed for some time as a way to learn and share insights. This process involved grandparents, parents, and children. The family made some exciting decisions that increased their cohesiveness, including construction of a log cabin that they all worked on at the location of a former cabin on their property. This became the family getaway for special events. The cabin is within the restored landscape on the land where the parents live, and provides a special place for all members of the family to gather.

Doing the work on the land is often more fun than preparing budgets. Because the fun tasks cannot reliably happen without funds, budgets are necessary. Also, tracking expenditures to make sure a project does not go over the allocated budget is often absolutely essential.

Budgeting Strategies for Restoration

There are different types of costs and cost estimates. Budgeting typically addresses four elements of anticipated costs: (1) tasks, (2) materials, (3) a contingency to incorporate some acceptable uncertainty around the estimated costs, and (4) the cash flow needs and timing, as to when the money would need to become available, aligned with the restoration schedule.

If you lack funding for labor for follow-up treatments, such as herbiciding stumps after brush is cut, the task can fail or you may encounter substantially more costs later on. Thus, understanding the whole budget, whether you do the work yourself or hire it done, becomes critical to the overall success of the restoration program. If you plan to initiate a task, make sure you budget for necessary follow-up work, including labor and materials to complete it.

If money is no impediment to undertaking restoration, we still encourage you to develop annual budgets and an overall program estimate (see step 7) for other important reasons. First, even where money has not been an issue, families and support often change, and newcomers to the project may have concerns about cost. Being in a position to "show and tell" the budget details is very important for communication and continuity purposes. Second, refining procedures as you learn often leads to refinement in budgets. Take full advantage of test and demonstration plots. Preliminary cost estimates are generally more accurate when based on small-scale test and demonstration plots.

Task 17. Prepare a Preliminary Budget

Here, we suggest you develop your preliminary budget. In the goals and objectives data form (data form 4.1, appendix 1), assign primary restoration tasks to achieve each objective (see table 4.2, as an example). Too often people embark on a restoration project with a focus on one task and fail to grasp the big picture. The restoration and management planning data form (data form 1.1, appendix 1) has been designed to display all commitments and demands of time and money, for the entire project. You do not have enough information yet to do a final budget, but you can get started.

Start your budgeting as you begin to define goals and objectives, and then refine the budget as the restoration plan develops, initially using the goals and objectives data form (data form 4.1, appendix 1) expanded to include a listing of primary restoration tasks (table 4.2). Then use the master budget data form (data form 4.2, appendix 1) to help you organize, and begin compiling and filling in costs and assumptions as you proceed through each step in the restoration planning process. Because this is only the first iteration of a continuing budgeting process, we call this a preliminary budget. It will later be refined, and ultimately become a final budget. To use the master budget data form, simply start replacing the types of ecosystems to be restored in the data form with the types of ecosystems to be restored in your plan (see table 4.3, as an example). As the acres of each type of ecosystem to be restored and the tasks required to restore them become better understood, continue to fill in the master budget data form. Include a contingency cost for uncertainty on the unit costs for each task. If you also need to purchase equipment, you can insert the prorated cost per acre associated with its purchase and operational costs. In other words, start developing per-acre unit costs for each treatments and task; this also will be refined over time as the plan and costs become clearer.

Depending on the size of the project and your finances, decide how you want to approach estimates of the cost of the project. The language surrounding budgets can be confusing depending who the budget is for, the complexity of the project, and the length of time the budget covers. All budgets are simply attempts to lay out and organize costs as clearly and accurately as possible, but sometimes greater accuracy is demanded. The following are budgeting strategies commonly used in the restoration industry:

TABLE 4.2.

Goals and objectives aligned with primary restoration tasks and mapped locations for treatments

Stressors	Goals	Measurable Objectives	Primary Restoration Tasks

Existing erosion
  Location A
  Location B
 	
  Location C
	Stabilize soil against erosion	100% rill of erosion stopped
		100% of gulley erosion stopped
			Mulch area with wood chips
			Divert trail/road runoff to new location, install several log water bars and replant
			Plant cover crop to provide > 90% quick-growing vegetation cover

Stormwater runoff and contamination
 	
  Location A
 	
  Location B
  Location C
	Intercept and prevent contaminated water from entering pond, wetland or stream	All off-site water is pretreated at property line	
			
			Intercept/treat by sacrificial biofilter wetland

Crop production fields Location A Location B	Convert crop fields to native vegetation by seedings and planting	In 3 years, 50 native species prevail	 Plant native grassland seed mix in area A Seed cover crop to stabilize soils, then over 2 years, plant acorns, hickory nuts, hazel nuts, to restore prairie and savanna
Invasive plant areas Species A Species B	Control and reseed with native plant seeds to establish native communities	In 3 years, invasive species reduced to < 1% cover and 50 native species prevail	 Hand pull garlic mustard in mapped locations Hand cut, stump-treat multiflora rose in mapped locations
Plant on-site native seed nursery for rarest plant species	Collect local seed, propagate and create local seed source	In year 2, establish 3 acres capable of producing ~300 lbs. of pure live seed per year for the restoration areas	Establish a list of target species, collect seed, propagate plants, plant nursery. Harvest, dry, clean, and store seeds annually for direct seeding into restoration areas.

TABLE 4.2.

Continued

Stressors	Goals	Measurable Objectives	Primary Restoration Tasks
Species A			Plant $1/4$ ounce/acre in mapped receiving area
Species B			Plant 8 ounces/acre in mapped areas
Species C			Grow and plug plant clusters of 25 plants each into the locations mapped

- *Back-of-the-envelope budgets*—it is usu-ally impossible to do a good back-of-the-envelope budget for an ecological restoration project because there are so many tasks conducted over many years, and costs often change annually. We discourage such casual budgeting unless it is a small project with minimal financial concerns.

- *Rough budgets*—These are estimates for use among close associates. For example, very early in discussions with a contractor, you may want a rough idea of what the project might cost. That estimate could be developed following some rigorous method, but because it is so early in the planning, it will still be only a ballpark idea. We discourage the use of rough budgets as a basis for entering into contracts for services. Rough budgets typically do not include all assumptions such as labor costs or well-researched unit and material costs. This and the back-of-the-envelope budget are often considered working budgets because they are modified as the job progresses.

- *Professional estimates or probable cost opinions*—These are created by contractors hired to provide an estimate of the costs.

They often are the general contractor who will be overseeing the project. Property owners or managers should work very closely with them to ensure that their goals and objectives are driving the budget.

- *Fixed pricing estimates with a defined contingency*—This is typically a bid price provided by someone proposing to do restoration work for a project. The contingency simply means the budget preparer has decided to include a 5% to 10% overage above the summation of the actual budget they developed by adding up all projected costs. A contingency also is often included for multiyear prices to account for price escalation over time.

- *Final budgets*—These are comprehensive, with a line item for each task, so that all anticipated materials, equipment, subcontractors, labor, state and federal taxes, and any other costs are fully listed. These budgets also will contain the assumptions you have made for each item in the budget. Final budgets cannot be completed until restoration planning is nearly done. They can be produced by you, or by a general contractor you hire to oversee the project.

TABLE 4.3.

Working budget on master budget data form: Example

Stone Prairie Farm Restoration Working Budget
Budget prepared by S. Apfelbaum
Cost estimate
Restoration task

Restoration task	Unit	Quantity	Unit Cost	Cost ($)	Assumptions
A. Emergent wetland restoration in tiled fields					
1. Ditch backfilling	linear feet (lf)	200	1	$200.0	cubic yards (CY) assumes haul in and place
2. Wetland planting acreage (emergent)	each	3	1,000	$3,000.0	500 plants per acre @ $2/each
3. Wetland seeding acreage	ac	3	1,200.00	$3,600.0	Seed cost × acres
4. Site prep herbiciding: 1st application	ac	3	200	$600.0	Herbicide weed treatment cost × acres
5. Site prep herbiciding: 2nd application	ac	3	100	$300.0	Herbicide weed treatment cost × acres
6. Site prep herbiciding: 3rd application	ac	3	100	$300.0	Herbicide weed treatment cost × acres
7. Vernal pond excavation and planting	ac	0.5	2,400.00	$1,200.0	1600 CY/acre-foot × 6-inch depth excavation @ $3/CY
8. Site prep: scarification	ac	3	125	$375.0	Tillage costs × acres
9. Soil bioengineering	lf	200	40	$8,000.0	live stakes @$40/lf
B. Wet mesic prairie restoration					
1. Exotic species/brush removal	ac	1	1,500.00	$1,500.0	acres
2. Wetland planting acreage	each	1	$4.50	$4.50	200 plugs per acre @ $4.50/plug. installed
3. Wetland seeding acreage	ac	1	2,500.00	$2,500.0	Seed cost × acres
4. Site prep herbiciding: 1st application	ac	1	200	$200.0	Herbicide weed treatment cost × acres
5. Site prep herbiciding: 2nd application	ac	1	100	$100.0	Herbicide weed treatment cost × acres
6. Site prep herbiciding: 3rd application	ac	1	100	$100.0	Herbicide weed treatment cost × acres
7. Site prep burn	ac	0	175	$0.0	See larger burn task
8. Site prep:scarification	ac	1	80	$80.0	Tillage costs × acres
C. Mesic and dry prairie planting					
1. Prairie planting acreage	each	6000	4.5	$27,000.0	100 plugs per acre
2. Prairie seeding acreage	ac	60	800.00	$48,000.0	Seed cost × acres
3. Site prep herbiciding: 1st application	ac	60	100	$6,000.0	Herbicide weed treatment cost × acres
4. Site prep herbiciding: 2nd application	ac	60	100	$6,000.0	Herbicide weed treatment cost × acres
5. Site prep herbiciding: 3rd application	ac	60	100	$6,000.0	Herbicide weed treatment cost × acres
6. Site prep burn	ac	0	175	$0.0	See larger burn budget line item
7. Site prep: scarification	ac	60	50	$3,000.0	Tillage costs × acres

TABLE 4.3.

Continued

Stone Prairie Farm Restoration Working Budget
Budget prepared by S. Apfelbaum
Cost estimate

Restoration task	Unit	Quantity	Unit Cost	Cost ($)	Assumptions
D. Savanna restoration areas					
1. Tile disablement	lf	100	4	$400.0	Assumes $/lf for disablement
2. Savanna shrub/tree planting	each	3750	5.00	$18,750.0	200 trees, 50 shrubs per acre @ $5 each installed, as seedlings
3. Savanna seeding acreage	ac	15	200.00	$3,000.0	Seed cost × acres: assume 8 lbs. per acre of native savanna grass and 1 lb. of forbs per acre
4. Site prep herbiciding: 1st application	ac	15	100	$1,500.0	Herbicide weed treatment cost × acres
5. Site prep herbiciding: 2nd application	ac	60	100	$6,000.0	Herbicide weed treatment cost × acres
6. Site prep herbiciding: 3rd application	ac	60	100	$6,000.0	Herbicide weed treatment cost × acres
7. Site prep burn	ac	0	175	$0.0	See larger burn budget line item
8. Site prep: scarification	ac	60	50	$3,000.0	Tillage costs × acres
E. Riparian forest buffer planting					
1. Riparian shrub/tree planting	each	250	5.00	$1,250.0	200 trees, 50 shrubs per acre @ $5 each, installed as seedlings
2. Riparian seeding acreage	ac	1	3,500.00	$3,500.0	Seed cost × acres: assume 8 lbs. per acre of native savanna grass and 1 lb. of forbs per acre
3. Site prep herbiciding: 1st application	ac	5	100	$500.0	Herbicide weed treatment cost × acres
4. Site prep herbiciding: 2nd application	ac	60	100	$6,000.0	Herbicide weed treatment cost × acres
5. Site prep herbiciding: 3rd application	ac	60	100	$6,000.0	Herbicide weed treatment cost × acres
6. Site prep burn	ac	0	175	$0.0	See larger burn budget line item
7. Site prep: scarification	ac	60	50	$3,000.0	Tillage costs × acres
G. Miscellaneous					
1. Erosion control	ac	80	50	$4,000.0	Planting barley @ 60 lbs. per acre, needed if restoration cannot immediately follow soil scarification

H. Maintenance activities for 5 years

1. Mowing approx. 60 acres	trips	4	3,000.00	$12,000.0	Assume $50/acre @ 60 acres
2. Herbiciding (10 acres)	trips	4	100.00	$400.0	assumes $100/acre @ 1.5 qts. Roundup/acre needed on up to 10 acres
3. Prescribed burn	each	2	250.00	$500.0	Assumes 80 acres each burn @ 5 hours @ $500/hr burning of entire site 2 times 5 years
4. Monitoring hydrology, vegetation for 5 years		1	5,000.00	$5,000.0	
5. Disposal of project site debris/garbage		1	250	$250.0	Pick up old farm debris and haul to dump: recycle
6. Final engineering and site design		1	5,000.00	$5,000.0	Finalizing culvert and ditch backfills
7. Permitting		1	2,500.00	$2,500.0	Assumed costs for permitting
8. Neighborhood interaction		1	500.00	$500.0	2 gatherings on farm for neighbors
9. Permit fees		1	500.00	$500.0	Assumed permit fees for submittal of application to county
Subtotal				$184,109.5	
Performance bonds ($20–$1000 Dollars)				$3,682.2	If required
Project administration (2.0% of subtotal)				$3,682.2	If required
Endowment $ (3% of total)				$5,523.3	
Contingency 20%				$36,821.9	Adjust as necessary for confidence
Grand total				$233,819.1	

Especially in more complete budgets or a final budget, where contingencies are included, you want to include worst-case and best-case estimates to provide sideboards on your anticipated costs. Worst-case estimates essentially capture the highest reasonable potential cost for each task. Best-guess estimates assume the best possible outcome of the uncertainties surrounding costs.

Keep in mind that budgeting complex restoration projects must be an iterative process. You typically will start by developing a back-of-the-envelope budget, then refine it to a rough budget, then move to a firm budget with a defined contingency, or accept professional estimates that are probable costs. A completed plan will contain a final budget. With the right tools to help you think more comprehensively about the costs in a project, you can more quickly develop an accurate budget. This should always be your goal.

In addition to the total cost estimate, budgets generally are more useful if they are broken down (1) by task, (2) by labor and material costs, and (3) according to a restoration schedule, so you will know the cash flow needs for your project. Cash flow is the timing of money needed to do the restoration work.

Contractors or Do It Yourself?

A good way to get an accurate budget for your project is to work with contractors and ask them for a bid for tasks that you cannot or choose not to do. To get bids, you need either clearly written, described, and mapped tasks or a scope of work that contractors can respond to by giving you an accurate bid, or, you can invite contractors to your land, and give them enough information that they can develop an accurate bid. The former is better unless your project or task is very straightforward—like mow and bale vegetation

in a particular field. Even mowing can involve details that may be important in restoration. For example, the mowing may need to be done at a time when you can herbicide fresh green regrowth. Always make sure there is agreement on the timing you need for when tasks must be completed, and get it in writing so there is no misunderstanding.

To get the best and most accurate bids, it is usually best to have the where, what, when, and how mapped and in writing in a bid package, to give to each bidder. A standard bid package will ensure this process delivers bids that have the highest value and accuracy, so you get good numbers and results. Obtaining bids from contractors does not obligate you to accept them. However, do not ask contractors to put the time into developing bids simply to generate your budget. Using contractors simply to get ideas and bid numbers will work only once.

There is another way to work with contractors, which does not take much of their time and allows you some flexibility in deciding how to get your work done. More often than not, contractors have their own internal spreadsheets with unit costs to do each task on a restoration project (see table 4.4). If you list the tasks you need, and the timing and acreages of each task, you can ask contractors for approximate unit cost ranges. Be sure to let them know you are using the estimates for your project budget. Their project bids are not generally shared with other contractors, but if you give them your short list of project tasks, they may give you the unit price ranges so you can more accurately develop your own budget.

Task 18. Refine Goals and Objectives

Once you begin laying out your goals and objectives, you will normally find yourself wondering

TABLE 4.4.

Typical unit rate table developed and used by contractors

Mobilization fees	
Price range	$150–$1,000 (or more)
Seeding	
Total cost range	$1,000–$4,000 per acre
Base material costs	$300–$1,700 per acre
Application method	$200–$1,000 per acre: labor, equipment, mobilization
Straw mulch	$700–$1,800 per acre: labor, equipment, mobilization
Site preparation	
Herbicide	$100–$1,000 per acre (tractor or ATV)
Tilling or disking	$100–$500 per acre
Maintenance	
Total cost range:	$800–$1,500 per acre (per trip)
Herbicide	$200–$1,000 per acre (per trip)
Prescribed burning	$460 per hour for crew of 4, plus mobilization & permitting
Mowing	$125–$300 per acre
Brushing	$2,500–$9,000 per acre: labor and materials
Planting	
2″ plug	$350–$7 each: labor & materials
Gallon	$10–$15 each: labor & materials
Trees	$45–$800 each: labor & materials (3″ diameter)
Shrubs	$55–$100 each: labor & root-pruned materials (to container size)
Water lilies	$12 each: labor & materials
Herbivore/wildlife protection fencing	$3–$8 linear foot: installation & removal
Erosion control and stabilizers	
Erosion control mat—temporary	$150–$550 square yard: labor & materials
Erosion control mat—permanent	$4–$7 square yard: labor & materials
Biologs	$35–$55 linear foot
Dead brush ascines	$7–$15 linear foot
Waterbars	$5–$12 linear foot
Stream or ravine restoration	$250–$550 per linear foot
Miscellaneous Items	
Silt fence	$250–$375 linear foot
Rain garden	$10–$25 per square foot

what it will all cost. You may find during the budgeting exercise that you cannot afford to do some of what you had hoped to accomplish. This task directs you to refine your goals and objectives by reviewing areas of uncertainty and by understanding regulations and other external constraints discussed earlier. This task also helps you soul search to more fully appreciate the personal and financial commitments you are making through the goals and objectives for your project. You should refine your written goals and objectives as you think through each of the issues.

Summary

Laying strong foundation stones is critical for the success of your restoration plan. Too often, projects fail not because the restoration work was too hard, but because not enough time was taken to address the human factors of leadership, decision making, and participation. Before beginning a restoration plan, be sure to carefully assess the leadership issues, and include all participants as much as possible in the development of goals and objectives as well as regular reviews. Inclusiveness is often imperative. People are invariably sensitive to changes imposed upon them, especially without their involvement in formulating the ideas and the program.

When thinking about goals and objectives, think first what you hope to achieve, given the ecological capability of your land. Then, consider the externalities of what your time and budget will permit. Final budgeting has to come after you have a good draft of your plan, after the next step, but it should begin as your plan is initially conceived, starting with defining the most basic goals and objectives. The preliminary budget you have developed should provide you with a good ballpark estimate of your costs and the cash flow needed in at least the first few years of the project.

Step 5.

Develop Your Restoration Plan

Unless commitment is made, there are only promises and hope; but no plans.

Peter F. Drucker

The best plans are those that are easily communicated to others. This is especially important where the restoration is a group effort. The plan should allow others to quickly find the right tables or figures with schedules, roles and responsibilities, material lists, and record-keeping forms. Our restoration plans have been evolving away from expository plans to digital and graphic plans, as described in this step. We want to emphasize, however, that this in no way reflects the importance of acquiring all useful background information on your project.

We recommend that restoration plans focus on what is essential and necessary to implement the restoration. There is no end to details that might be included if the plan were an academic exercise, but your aim should be to produce a pragmatic plan that makes interpretation as easy as possible. There usually will be background documents and information with useful details. Put them on your library shelf where you can find them as needed, but do not burden the restoration plan with them.

Types of Restoration Plans

Expository plans are commonly one to two hundred pages with many attached appendixes. Graphic restoration plans, which we recommend, present virtually the entire plan with tables, aerial photographs, and maps. Everything, including goals and objectives, treatments, methods, and schedules, is configured to fit into tables and integrated maps similar to what you have already prepared. There is a separate page for each topic, with all information needed on that subject condensed into the single page. This makes it quite easy to stay organized and not get overwhelmed by background information and extraneous details. For example, it is very convenient to go to a monitoring page for an ecological unit to get all needed details of the layout, techniques, schedules, and responsibilities for the monitoring work without having to dig through pages of information. Restoration activities are all captured on the maps of each management zone or unit, with tabular instructions on what to do in each. The other advantage of a graphic restoration plan is that you can record your work in each unit each day, what was actually accomplished, hours, expenses, and other important records. This can be done either with hard copy or with digital records.

Graphic restoration plans facilitate communications. At meetings, you can lay out plan sheets on a table and talk through each one. Others can easily see the geography of the restoration program in the maps and aerial photographs. Graphic plans also are easily converted into work orders for contractors or volunteers, to show them where to go to implement a particular task. You can more easily convert the graphic restoration plan to presentations for a larger audience.

If for some reason you want or need an expository plan, it is easy to develop by explaining each component of the graphic plan and appending maps, tables, and reference materials. Both types of plans can be generated in digital form, although graphic plans require larger digital memory. For this reason, they are more easily distributed as a CD-ROM. CDs provide good back-up to computer files, but must be updated as plans evolve.

Task 19. Design the Outline (Content) of Your Plan

The organization of this workbook is how you should organize your restoration plan, although depending on the complexity and scope of your project, you may be able to skip over some subheadings or tasks. Use the restoration and management planning data form (data form 1.1, appendix 1) to check the contents you may wish to include in your plan.

About half of what you should include in your plan you have already completed but should review and revise as necessary. We provide a summary of the important items that should go into your plan in the balance of this step. Items not yet discussed are covered in the remaining steps (steps 6 through 10). As you complete the remaining tasks, insert the information into the plan you begin to develop now. These deferred tasks are no less important, but must be undertaken after you have a good draft of your plan.

Guiding Principles

The guiding principles for a restoration program should be inspiring and informative. They are your vision of what you hope to achieve. We have deferred discussion on these principles until now because until you could begin to understand what was possible in your restoration, it was futile to spend time on a vision for it. In the case of Stone Prairie Farm, the guiding principles were developed to help parents communicate their intentions to their children who would someday take over responsibility for the restoration plan.

1. Stable soils will be present on the land.
2. Cold and clear waters will run in the stream.
3. Native plants in appropriate numbers and patterns will be growing on the land again.
4. Desirable native wildlife will again become plentiful on the land.
5. Stone Prairie Farm will be appreciated as an important place to learn about healthy ecological systems.
6. We will share with others what we learn from this land.

Basemap

You developed and refined a basemap at two scales in step 1. With what you now know about the information associated with the basemaps and how they are used, are these scales still appropriate? If not, revise and redraw the two basemaps. See figure 5.1 as an example of how your basemap should look.

Refine the existing conditions or vegetation cover maps for your project area. Use as few maps as possible, although it is common to have a separate map for each ecological unit, especially if they are large or variable. Figures 5.2 and 5.3 are examples of the existing conditions maps.

Historic Conditions Maps

You prepared historic conditions or vegetation cover maps in step 2. These maps show your best guesses of how the property looked before it was altered by people. This information underpins your understanding of the stressors and changes that have occurred in your property. Because the maps you developed are based on your

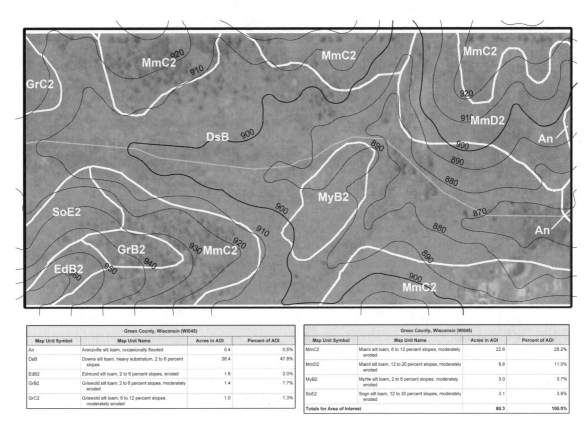

Green County, Wisconsin (WI045)			
Map Unit Symbol	Map Unit Name	Acres in AOI	Percent of AOI
An	Arenzville silt loam, occasionally flooded	0.4	0.5%
DsB	Downs silt loam, heavy substratum, 2 to 6 percent slopes	38.4	47.8%
EdB2	Edmund silt loam, 2 to 6 percent slopes, eroded	1.6	2.0%
GrB2	Griswold silt loam, 2 to 6 percent slopes, moderately eroded	1.4	1.7%
GrC2	Griswold silt loam, 6 to 12 percent slopes, moderately eroded	1.0	1.3%

Green County, Wisconsin (WI045)			
Map Unit Symbol	Map Unit Name	Acres in AOI	Percent of AOI
MmC2	Miami silt loam, 6 to 12 percent slopes, moderately eroded	22.6	28.2%
MmD2	Miami silt loam, 12 to 20 percent slopes, moderately eroded	8.8	11.0%
MyB2	Myrtle silt loam, 2 to 6 percent slopes, moderately eroded	3.0	3.7%
SoE2	Sogn silt loam, 12 to 30 percent slopes, moderately eroded	3.1	3.8%
Totals for Area of Interest		80.3	100.0%

FIGURE 5.1. Basemap with soils and topography, Stone Prairie Farm, WI

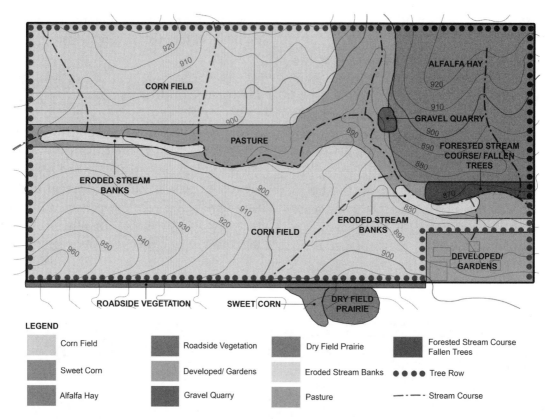

FIGURE 5.2. Existing conditions assessment for Stone Prairie Farm, WI

understanding of historic conditions, they probably require no revision unless you have gained additional insight into the past since you completed them. See the following examples (fig. 5.4).

Summary of Stressors and Resulting Changes

Review the graphic summary of the stressors and change analysis for your project site that you completed in step 3. If you have not done so, map the stressors that currently operate, depicting each with a map symbol so it is clear and distinct,

as in figure 5.5. It is especially useful to match these as best you can to corresponding healthy reference areas on your property or in nearby natural areas (fig. 5.6). Develop summary maps depicting the locations on your property that have changed (refer back to figs. 1.2 and 1.3 for examples). Include the stressors and changes that are not believed to be addressable at this time. These include changes or stressors resulting from conditions beyond your property as well as conditions on your property that are beyond your ability to correct, at least at this time. An example might be serious down-cutting of a stream or overabundance of deer or elk. You probably will want to prepare a narrative to go with the maps explain-

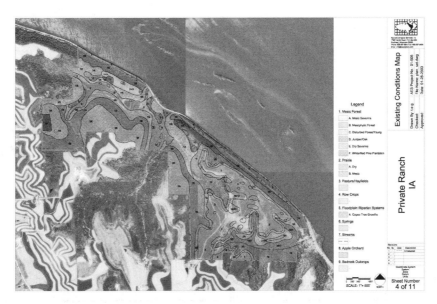

FIGURE 5.3. Existing conditions assessment for a private ranch, IA

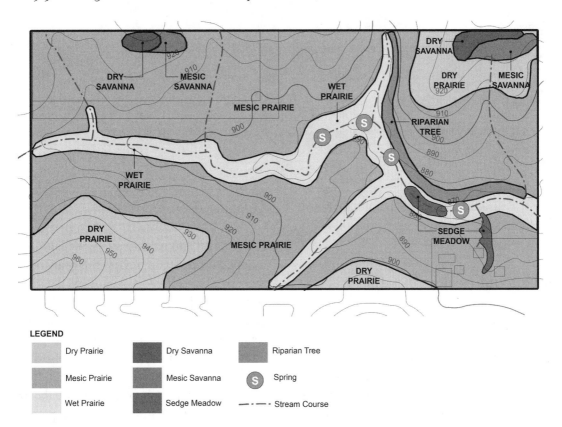

FIGURE 5.4. Historic condition assessment for Stone Prairie Farm, WI

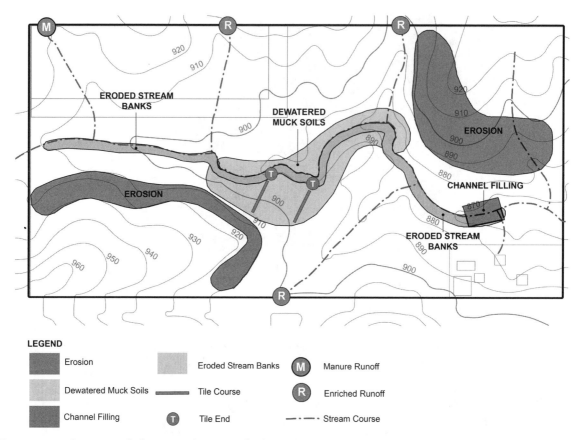

FIGURE 5.5. Stressor and change analysis map for Stone Prairie Farm, WI

ing stressors and hypothesized changes that have resulted, but keep it as brief as possible.

Gradient and Ecotone Analysis

The gradient analyses completed in step 3 now need to be summarized and finalized. Figure 5.7 provides an example of a gradient analysis that can be used with reference sheets that describe the existing degraded and healthy conditions (fig. 5.6). Complete ecotone and gradient graphics and narratives on your project site.

Goals and Objectives

Especially if you have made modifications in the mapped information, review the goals and objectives you developed in step 4. Make sure your goals and objectives are consistent with the principles you stated earlier. As much as possible, state objectives that can be measured. We will say more about that in step 6 where monitoring is discussed. Historic and existing conditions, stressors, and the summary of changes are the primary pieces of information used in developing project goals and objectives (see table 4.3). Make sure

Oak Savanna Systems

Healthy Systems

General Structure

- Semi-open to open tree canopy
- Multiple age classes of trees
- Dominant cover of native grasses, sedges and forbs
- Natural oak regeneration
- Sporadic native shrub layer
- High light levels interspersed with partial/isolated shade

Indicator Species of Healthy System

- Bur oak
- White oak
- Black oak
- Shagbark hickory
- Savanna ground layer species

Soil Profile/Topography/Hydrology

- Well-drained silt, clay and sand loams, gravely sands, alluvium glacial features
- Higher and dry sites, and moist, well-drained soils

Associated Species

- Pennsylvania sedge
- Silky and Virginia wild rye
- Bottle grasses
- Other sedges
- American hazel shrub

Unhealthy Systems

General Structure

- Continuous, closed canopy
- Dense layer of non-native shrubs
- Bare, eroding soil
- Low light levels, predominant dense shade
- No oak regeneration
- Few or no young age classes of trees
- Lack of native ground cover vegetation
- Encroachment by development or agriculture

Indicator Species of Unhealthy System

- European buckhorn
- Tartarian honeysuckle
- Black locust
- Boxelder
- European brome, Kentucky bluegrass, and other non-native grasses
- Agricultural weed species and brambles

Protection and Management Considerations

Causes of Change

- Cessation of historic fire regimes
- Destruction due to urban development
- Invasion of competing non-native shrubs
- Encroachment of adjacent development with associated pollutants
- Intensive grazing
- Change in hydrologic regime (drier or wetter)

Restorative Capacity

- Highly restorable under well-designed and implemented restoration and management program
- Highly disturbed sites may require replanting of native species, especially ground cover, if native seed bank is absent

Protection Strategy

- Adopt land development practices that place a high priority on ecological protection beyond that of existing wetland ordinances
- Implement an annual, long-term restoration and management plan
- Protect historic hydrologic regime/systems

FIGURE 5.6. Healthy and unhealthy ecosystem examples

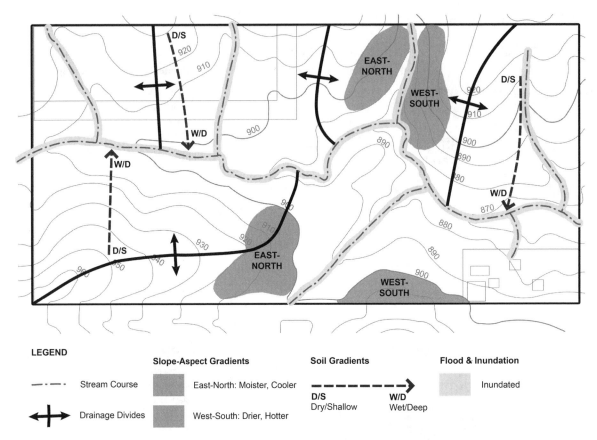

LEGEND

| Slope-Aspect Gradients | Soil Gradients | Flood & Inundation |

- · — · Stream Course
- ◄► Drainage Divides

East-North: Moister, Cooler

West-South: Drier, Hotter

D/S — ► W/D
Dry/Shallow   Wet/Deep

Inundated

FIGURE 5.7. Gradients, ecotone patterns, and ecosystem types at Stone Prairie Farm, WI

your goals and objectives are consistent with that information. If your land once supported deciduous forest, then your restoration should move ecosystems in that direction. In most cases, restoration goals should be compatible with where nature would like to go.

Targeted Ecosystems or Ecological Units

The next step uses what you have learned about the achievable ecological restoration opportunities on your project site. For this plan sheet you will want to consider the gradients, ecotones, and understanding of historic ecosystems, and their distribution on the project site. Given your goals and objectives, how will the cover types be ideally distributed on your project when restoration is completed? This might be seen as a graphic representation of your goals. This map should depict the types of ecosystems, plant communities, or ecological units that you hope to reestablish. You might even set some priorities for those most important to you. Examples of targeted ecosystem are included in figures 5.8 and 5.9.

Conservation Plan

A conservation plan deals with offsite stressors you previously identified and mapped in step 1

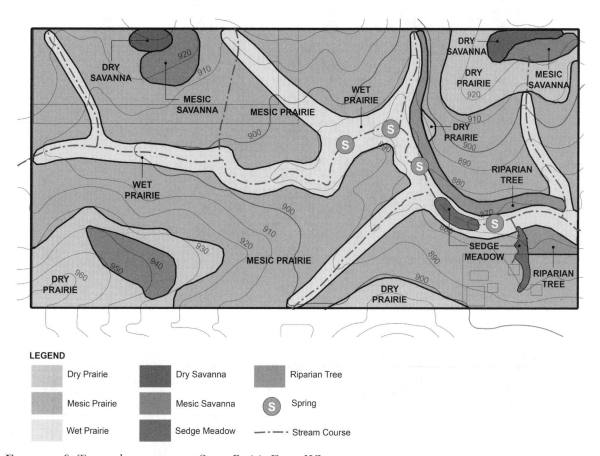

LEGEND

Dry Prairie	Dry Savanna	Riparian Tree
Mesic Prairie	Mesic Savanna	(S) Spring
Wet Prairie	Sedge Meadow	—·—·— Stream Course

FIGURE 5.8. Targeted ecosystems at Stone Prairie Farm, WI

(task 8). Sometimes it is possible to address offsite stressors by entering into conservation and land management agreements with neighbors, or by implementing strategies on your own project that are focused on protecting your land and the restoration investments you are making. Figures 5.10 and 5.11 provide examples of the conservation planning elements for Stone Prairie Farm and another project.

Notice that at Stone Prairie Farm, the conservation elements were focused on enhancing the quality of surface water entering the project from neighboring farm lands. This included the capture and cleansing of nutrient-enriched feedlot runoff and farmland runoff with vegetated biofilter wetlands. These were sized to contain the runoff and hold it for an adequate time to cleanse the water of sediments and excessive nitrogen, through denitrification by bacteria living in the wetland biofilters.

In the second example, the conservation plan aimed to secure a partnership with neighboring landowners to change uses on their land to affect improvements in the restoration project. The primary improvements buffered the rapid storm water runoff from steep neighboring agricultural fields. With neighbor cooperation, vegetated buffers were developed to restrain and filter

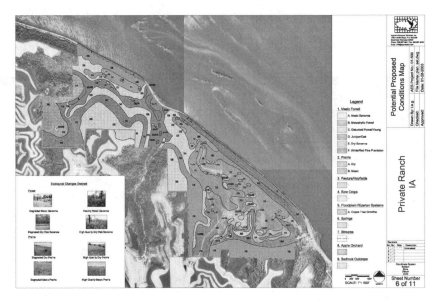

FIGURE 5.9. Targeted ecosystems at a private ranch, IA

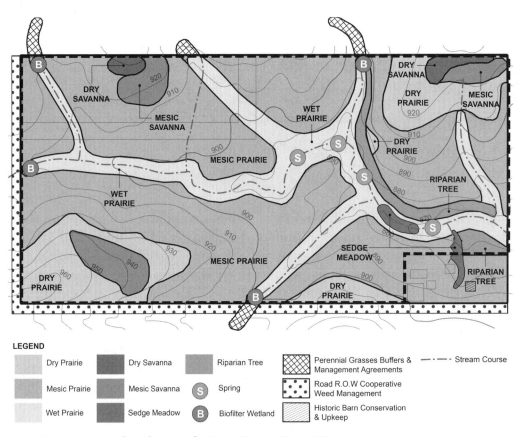

FIGURE 5.10. Conservation plan elements for Stone Prairie Farm, WI

FIGURE 5.11. Conservation plan elements for a private ranch, IA

storm water. The plan involved partnering and cooperation in restoration and management on each other's land.

Obviously, the conservation needs of each project will vary, but they consistently focus on ways to reduce or mitigate stressors affecting the project site. These might involve conservation easements with potential tax benefits for participants. Many elements of conservation planning are eligible for financial assistance through government programs, so check with your NRCS office or county conservationist.

Remedial Management Units

Define the boundaries of areas where you will focus common remedial restoration treatments. Management units are often different from ecological units (targeted ecosystems or ecological units). Think of them as areas where common treatments will be applied at the same time. They need not be contiguous; they might be labeled as management unit 1a, 1b, etc., if more than one ecological unit is combined into the same management unit. Be mindful that management units may need to be redefined over time. For example, you may initially identify areas needing the same preplanting preparation; areas receiving a different type of preparation would be mapped as a separate management unit. Both, however, might be seeded with the same seed mixture. Either create two management unit maps, one depicting preplanting treatments and the second the planting treatments, or as an alternative, you might map both as one unit, broken into subunits where different site preparations are to be applied. After restoration treatments give way to maintenance management, usually many units can be combined into the same management unit.

At Stone Prairie Farm, the management units were laid out to initially stabilize erosion. We began on upper slopes and progressively worked

FIGURE 5.12. Management units for Stone Prairie Farm, WI

toward the lower areas, including the stream course (see fig. 5.12). Any high-quality ecological ecosystem probably should be a separate management unit and planned for immediate restoration work, while lower quality, even if similar, might be separated and put into a lower priority. Remember the principle: restore the best areas first.

You may be able to also prepare maps showing maintenance restoration units if you can project five or so years ahead when remedial restoration has been completed (see fig. 5.13). If not, these can be done later. Be realistic about what you can accomplish. For example, if you were to use prescribed burning as a primary maintenance strat-

egy, burn units should reflect what you can safely undertake in an afternoon or day. In figure 5.13 these management units are defined primarily by trail access, which also serves as fire breaks.

The relationship between recreational trails and management units is important to achieving many ecological outcomes. For example, where trails are used, such as around the perimeter of a wetland, too often land managers mistakenly locate the trail between different ecosystem types breaking up the ecotone then perpetuate the error by treating them as different management units. The restoration plan in this case will progressively eliminate the ecotone between the sys-

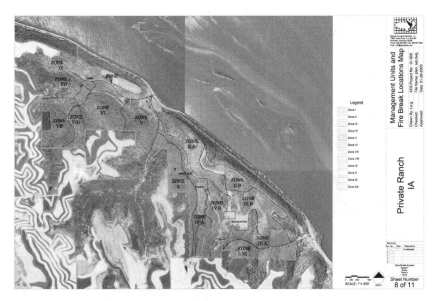

Figure 5.13. Management units for a private ranch, IA

tems. *Ecotones should be recognized and allowed to develop.* The disproportionately higher diversity typically found in an ecotone will be realized only if the ecotone is not fragmented. Avoid putting in trails or fire breaks parallel to gradients. Instead, as much as possible, design them perpendicular to gradients as depicted in figure 5.14. In this example, the two ecosystems might be treated separate for remedial work and then as one maintenance management unit, but the trail would not be located between them.

A Monitoring Plan

A complete restoration plan must include the details of monitoring for measuring success of your restoration. Details for a good monitoring plan are covered in step 6, so you will come back to this item later. Figure 5.15 depicts an example monitoring program using quantitative measurements with permanent sampling locations. Make

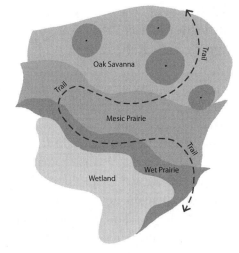

Figure 5.14. Careful design of trails ensure that ecotones are not compromised

sure that any monitoring you conduct will provide meaningful measures of actual changes in the land. The monitoring program is the basis for adaptively managing your restoration program as you move forward.

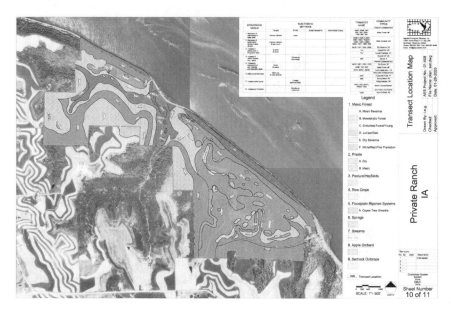

FIGURE 5.15. Monitoring plan for a private ranch, IA

An Implementation and Management Schedule

In step 7, you will develop schedules for each activity or task you listed previously, so you will come back later to complete this in your restoration planning worksheet. We provide examples in tables 5.1 and 5.2 for a Wisconsin nature preserve. Schedules are not set in stone. They will likely need to change with weather conditions, unforeseen budget or time constraints, and so on. It also is common to find that the demands of the on-the-ground restoration are different than expected.

Roles and Responsibilities

As you complete the previous items, give careful consideration to who will be responsible for doing the tasks and following the schedule you de-

velop. Determining who will be responsible to do the work often requires a two-level reality check. You should ask yourself the following questions:

- Will I or my family and committed friends have the time available to do the scheduled work when it needs to be done?
- Can I realistically start and complete tasks on the schedule I have developed?
- Have I honestly and clearly separated my ambitions from what is likely to be my true availability for doing the scheduled work?
- Do I need to adjust my ambitions and scale back the acreages to spread out the tasks to bring them into better alignment with my answers to the above questions?
- Which tasks need adjustments in scheduling to more realistically align the tasks with my availability to implement them?
- Are there ways to work with others, perhaps neighbors or volunteer groups, to address

TABLE 5.1.

Example of a five-year implementation, management, maintenance, and monitoring schedule, and estimated fees

	1995 QTR	1996 QTR	1997 QTR	1998 QTR	1999 QTR
1. Prescribed burn site inspection: Assess site conditions to determine feasibility, fuel load conditions	1 2 [3]* 4	1 [2] [3] 4	1 [2] [3] 4	1 [2] [3] 4	1 [2] [3] 4
Estimated cost:					
2. Burn management: Apply for burn permits, schedule burn, contact local authorities, finalize burn plan	1 [2] [3] 4	1 [2] 3 4	1 [2] 3 4	1 [2] 3 4	1 [2] 3 4
Estimated cost:	$100	$105	$110	$120	$130
3. Conduct burn: Up to 5 hrs. with a 2 hr. minimum at $280/ hour	1 [2] [3] 4	1 [2] [3] 4	1 [2] [3] 4	1 [2] [3] 4	1 [2] [3] 4
Estimated cost:	$1,400	$1,470	$1,540	$1,620	$1,700
4. Weed management and site inspection: Assess site condition, identify threats, for example, reed canary grass. Recommend mowing where necessary and/or design herbicide application plan.	1 [2] [3] 4	1 [2] [3] 4	1 [2] [3] 4	1 [2] [3] 4	1 [2] [3] 4
Estimated cost:					
5. Brush mowing: Conducted once annually	1 [2] [3] 4	1 [2] [3] 4	1 [2] [3] 4	1 [2] [3] 4	1 [2] [3] 4
Estimated cost:	$1,000	$1,000	$1,000	$1,000	$1,000
6. Herbicide management: Wick application to nonnative invasions, reed canary grass, and shrub invasion (up to 20 hours per year at $__/hr + materials)	1 [2] [3] 4	1 [2] [3] 4	1 [2] [3] 4	1 [2] [3] 4	1 [2] [3] 4
Estimated cost:	$2,000	$2,000	$2,000	$2,000	$2,000
7. Summary report: Annual report to provide specifics on activity and recommendations	1 2 3 [4]	1 2 3 [4]	1 2 3 [4]	1 2 3 [4]	1 2 3 [4]
Estimated cost:	$200	$210	$220	$230	$240
8. Vegetation monitoring: Biannual quantitative field sampling and report	1 2 3 [4]	1 2 3 [4]	1 2 3 [4]	1 2 3 [4]	1 2 3 [4]
Estimated cost:	$200	$210	$220	$230	$240
9. Hydrologic monitoring: Installation of water level recorder, and staff gages and piezometer.	1 2 3 [4]	1 2 3 [4]	1 2 3 [4]	1 2 3 [4]	1 2 3 [4]
Estimated Cost:	$2,046				
10. Hydrologic monitoring: Quarterly data retrieval and report.	1 2 3 [4]	1 2 3 [4]	1 2 3 [4]	1 2 3 [4]	1 2 3 [4]
Estimated Cost:	$1,386	$1,455	$1,527	$1,603	$1,683
Total Fees by Year	$10,382	$8,590	$8,847	$9,123	$9,403

Note: *[Bracket] indicates quarter when work will be conducted.

TABLE 5.2.

Restoration and management schedule with tasks

		Target natural communities				
Year	Restoration tasks	2A Wet prairie	2C Mesic Prairie	3B Mesic Savanna/oak opening	4A Sedge meadow	Total Acreage
1	Burn	0.00	0.00	12.50	4.10	26.10
1	Brush Removal/thinning	0.00	0.00	5.00	0.00	10.50
1	Weed control	0.00	0.00	5.00	4.10	14.60
1	Seeding/planting,	0.00	0.00	5.00	0.00	5.00
	remedial enhancement	0.00	0.00	0.00	0.00	5.50
1	Mow	0.00	0.00	0.00	0.00	0.00
1	Tile disablement	0.00	0.00	0.00	0.00	0.00
1	Grade controls	0.00	0.00	0.00	0.00	0.00
1	Monitoring & reporting	0.00	0.00	1.00	1.00	3.00
2	Burn	0.00	0.00	12.50	4.10	26.10
2	Brush removal/thinning	0.00	0.00	5.00	0.00	9.00
2	Weed control	0.00	0.00	10.00	4.10	23.60
2	Seeding/planting,	0.00	0.00	5.00	0.00	5.00
	remedial enhancement	0.00	0.00	0.00	0.00	4.00
2	Mow	0.00	0.00	0.00	0.00	0.00
2	Tile disablement	0.00	0.00	0.00	0.00	0.00
2	Grade controls	0.00	0.00	0.00	0.00	0.00
2	Monitoring & reporting	0.00	0.00	1.00	1.00	3.00
3	Burn	0.00	0.00	33.50	4.70	48.70
3	Brush removal/thinning	0.00	0.00	9.50	0.00	9.50
3	Weed control	0.00	0.00	19.50	4.70	33.90
3	Seeding/planting,	0.00	0.00	9.50	0.00	9.50
	remedial enhancement	0.00	0.00	0.00	4.10	6.30
3	Mow	0.00	0.00	0.00	0.00	0.00
3	Tile disablement	0.00	0.00	0.00	0.00	0.00
3	Grade controls	0.00	0.00	0.00	0.00	0.00
3	Monitoring & reporting	0.00	0.00	1.00	1.00	5.00

some tasks? Can this be managed for my project?

- If I choose to not make adjustments in the scheduled tasks, for whatever reasons, and doubt that I can find or trust volunteers to be available when I need them, can I hire help with those tasks?

At Stone Prairie Farm, Steve hired professionals to do the herbiciding and prescribed burning work. He made this decision so he didn't have to buy and maintain herbicide application equipment (e.g., tanks, pumps, ATV) and become engrossed in herbicide management, and to ensure that fires were safely administered. (See

Apfelbaum, Steven I. "Playing with Fire" in *Nature's Second Chance*. Boston: Beacon Press, 2009. to understand why Steve hired help to do prescribed burning.)

- Are there tasks that local farmers or high school students can help with? Are there local colleges or other organizations, such as job training programs, that might use your projects as a training opportunity?

Steve traded with neighboring farmers to get site preparation and old fence rows removed.

Complete Preliminary Budgets

You were first introduced to the budgeting process in step 4 where you prepared a preliminary budget (task 17). Now is the time to review the budget, making adjustments as the acreages of management units and scheduling of tasks is better defined. You may find that some of your earlier assumptions were incorrect, and you need to adjust the budget accordingly. For example, as you considered roles and responsibilities, it may have become clear that you will need to hire help or contract for some projects. Later, we will discuss other financial matters that will prepare you to develop a final budget.

Record Keeping

Keeping good records of what, when, where, and who did the work is very important (see step 8). You not only learn a lot about how ecosystems are responding to treatments, but also gain insight into costs, labor, seasonal timing, and other aspects that allow you to become more efficient and more successful in time. In short, good records lead to continuing refinement of the restoration program and are the basis for communicating project details to others. The records you keep not only help ensure the success of your restoration, they also enrich the experience for you and others involved in your restoration.

Appendixes

Appendixes can be attached to the graphic restoration plan. This is the place to put any detailed specifications for your project. For example, if you are going to hire labor to work on some tasks, getting bids for this work can require a clear definition of what the work entails and soliciting the bids in a standard format. We include an example of contractor qualifications and examples of construction contracts (appendix 3). In the website for *Restoring Ecological Health to Your Land*, you can download other contracts designed for hiring the services of professional restoration contractors, or to communicate intentions to others working on your restoration program.

Final Project Budget

Good estimates of all costs of your restoration are important for at least three reasons: (1) to develop a workable budget to fit with your personal finances, (2) to have estimates for negotiation with outside vendors for their services, or to barter for services, and (3) to have good estimates to support applications for grants. With the completion of this planning, you will be ready to do a final budget. We cover the remaining details in step 7 where you are directed to complete a final budget (task 27). Costs to implement restoration tasks are typically highest during the remedial restoration phase and much lower during the continuing maintenance restoration phase, so expect the cost to decrease greatly after the first three or four years.

A Perpetual Maintenance Budget and Endowment

We included an endowment line item in the master budget data form (data form 4.2, appendix 1) to help you think about long-term financing. This subject deserves careful attention. Too often, people do not consider how to finance the long-term (perpetual) maintenance of their restored land. This can become a serious issue with your heirs or assigns. The best way to ensure that your accomplishments have no long-term benefit is to ignore maintenance funding in your planning. Because it is difficult to get maintenance dollars through public budgeting processes, restored public lands often become neglected and fall into poor ecological health. You can avoid this on your own property if you plan ahead. We cover the details of developing the maintenance budget and endowment in step 7, task 27.

Summary

Good budget estimates require that you reach the point in your planning where you can see the specific tasks and what is necessary to accomplish them. The tasks, in turn, depend on your goals and objectives, which depend on the ecological capability of the land. Because all these things must be compatible and come together in the final plan, it is important to review them as the plan is developed and new insights and information become available. Even as the plan is implemented, you will continue to gain additional insight and may periodically want to revisit not only early assumptions, but even goals and objectives. Budgeting is an important process to determine what you can realistically achieve. Budgets for restoration projects require careful attention to details and recognition that needs change over time. The exercise of creating and maintaining budgets is often underappreciated. However, if you have limited funding or concern over the availability of adequate funding when you need it (cash flow), including future maintenance needs, budgets must be considered from the very beginning. We strongly encourage you to create and maintain accurate project budgets. This also includes record keeping on the actual costs you incur as you progressively implement your project.

A budget must be somewhat dynamic. Changes should be expected. Continually updating the budget estimate with the actual costs you experience through the process of implementing your plan is the best way to ensure your project does not spin out of control. If at all possible, you should also think about funding for the perpetual maintenance of your restoration. It is often unrealistic to assume that heirs will have the same priorities for the health of your land.

Develop a Good Monitoring Program

The only thing that is constant is change.

Heraclitus

Monitoring and good records are key to knowing how restoration treatments are altering the ecosystem and understanding the changes. The more complex the system, or the more carefully you wish to manage it, the more elaborate monitoring and records must be. Pilots are taught to take monitoring very seriously and keep careful records; dozens of conditions are continually monitored on a large plane. It is no less essential to monitor ecosystems. Although ecosystems are infinitely more complex than a commercial airliner, monitoring and good records are often ignored, or taken too casually. Here we provide guidelines, sources of information, and forms to assist in monitoring.

One might be tempted to point out that nature flies without a pilot, that ecosystems lack gauges and operate without a set of instructions. Indeed, if your management plan calls for hands off, and you are willing to let natural processes go where they will, monitoring and records are unnecessary, but don't confuse that with ecological restoration. You would never fly a plane that way (at least not for long), nor should you attempt to restore or manage an ecosystem with such a laissez faire attitude. By engaging in active manage-

ment, whether in the practice of forestry, fisheries, agriculture, or restoration, you are manipulating ecosystems, often profoundly. Monitoring is necessary to ensure that you are doing so in a sustainable and positive way. Monitoring is ideally designed to be a data collection process that provides specific response information, commonly compared to pretreatment baseline conditions, but even qualitative monitoring can provide much useful information, as we will discuss.

Task 20. Develop Your Monitoring Program

Monitoring should be carefully planned. It is most useful when conducted systematically over two or more years, ideally over the length of the restoration project. Keep in mind that as ecosystems respond during restoration, the kind and frequency of monitoring will change, nearly always becoming less intensive. Do not, however, initiate more than you can sustain. It is better to have long-term consistent data of a few important parameters than inconsistent or short-term data for many. To be useful, observations and data must be religiously recorded, with dates and locations. What may seem trivial might provide insight over time and lead to a more successful restoration.

To develop a monitoring program:

1. Review the guiding principles (task 19) to refresh your memory of what you hope to achieve with your restoration.

2. Consider one management unit at a time and review the goals and objectives you have for that unit (step 4).

3. For each goal and objective, list the observable and/or measurable changes that would provide convincing evidence that the ecosystem is moving in the desired direction or that the goal and objective have been achieved.

4. After reviewing the information in the remainder of this step, decide for each goal and objective how you will monitor to provide the observable and/or measurable information needed for steps 3 and 4. You might want to include at least preliminary ideas on how you will scale back the proposed monitoring for each unit as ecological health is recovered. You will need to include monitoring in your annual review (step 9) and make adjustments over time.

5. Compile monitoring plans, with details of methodology, for each management unit.

Formal monitoring can be either qualitative or quantitative, but systematic monitoring is essential, even if only qualitative observations are made. In most projects, monitoring is required for the following and should be designed considering the typical phases of a restoration project (see fig. 6.1).

1. *Baseline data gathering.* In the process of developing a management plan, it is necessary to assess the ecological condition and identify stressors before treatments are initiated. Using a systematic, standardized approach to baseline data gathering not only better ensures that you have examined critical aspects of the ecosystem, but also provides the baseline against which you can compare responses after treatment has begun. Detection of change is the primary aim of ecological monitoring.

2. *Annual monitoring to guide the process of restoration.* The kinds and amount of monitoring are dictated by the size and scope of the restoration project. Minimal monitoring may be sufficient for small, simple projects, such as a few acres of prairie or woodland, where periodic visual observation during the growing season may suffice. Monitoring is always required to detect new stressors or change in previously identified stressors. For example, all land, regardless of size, should be routinely and systematically examined for invasive species. If the land is located where environmental insults from neighboring property is likely, that, too, should be assessed annually. After initiation of treatments, annual monitoring is advisable for several years to ensure that the system is responding as planned.

3. *Project-specific monitoring.* In larger, more complicated restoration projects, some experimentation often is needed, usually with test plots. Which treatment provides the best results? Is there better control of an invasive species if burning is done in early summer rather than in early spring? Experimentation (discussed later) is a more formal way to allow nature to inform you of the best approach to restoration, and to get a better idea of labor and costs. This monitoring may need to be repeated for several years, at least until the efficacy of treatment is resolved.

4. *Long-term maintenance monitoring.* As initial goals and objectives are realized, remedial management gives way to maintenance with a corresponding decrease in time and cost. Likewise, monitoring can be reduced to a minimum,

but not neglected. On simple projects, this superficial monitoring might be accomplished by periodic walks on a Sunday afternoon. It is a continuation of the love affair between you and your land. The more familiar you are with the land (broadly defined, i.e., plants, animals, hydrology, etc.), the more easily you will detect changes that may indicate new problems. In more complex projects, long-term monitoring often involves data collection, but perhaps only every few years in each management unit. Both project-specific monitoring and long-term monitoring can help verify progress toward goals and objectives.

Experimentation/Test Plots

When you encounter questions about efficacy of treatments, consider small test plots to determine the best course of action. With good records of time and expense, small test plots also can provide important data on labor and costs for application of treatments.

Look for an area within the proposed treatment unit that is as representative but as homogenous as possible. Alternatively, test plots can be placed elsewhere, as long as conditions are similar to the proposed treatment unit. The size of the test plot depends on the treatment being tested. If you want to determine how well certain species grow, or how best to prepare the seedbed, a few square yards may be sufficient. If you want to determine if spring or autumn prescribed fire is best for controlling an invasive species, you should use plots at least a quarter-acre or larger. The most useful data will be obtained by comparing two or more treatments (see fig. 4.3). One treatment may be a control, although that often is provided by the entire management unit you intend to treat. State the question(s) as clearly as possible. For example, *Does autumn fire have a*

greater negative effect on species X than spring fire? Scientists prefer to state research questions as a null hypotheses: *There will be no difference between spring and autumn fires in the effect on species X.* Or, if using a control: *There will be no differences between spring and autumn fires and the unburned control in the effect on species X.* Decide at the outset how you intend to measure the effect on species X. Perhaps it will be total cover, number of stems, total fresh weight per unit area, or some other measure. We will discuss how these data should be collected later. Replication is always a good idea, and is achieved by putting in two or more test plots of each treatment.

Deciding What to Monitor

Because a primary aspect of restoration is mitigation of stressors, direct measurement of them is good if it can be done without extraordinary cost and effort. Otherwise, indirect indicators of changes in stressors, including decrease of invasive species and increase in desirable species, can be monitored. Ecological processes are generally difficult to monitor directly without sophisticated instruments and techniques but can be inferred from composition and structure. The organisms (composition) not only reflect the physical and chemical environment, but are responsible for the ecological processes and ecosystem structure that restoration aims to restore. When monitoring biota, always think first of keystone species—those that have the greatest impact on the overall ecosystem.

What you decide to monitor as well as what you need to confirm progress toward your goals and objectives determine how you monitor. Each of the following will require different measurement strategies:

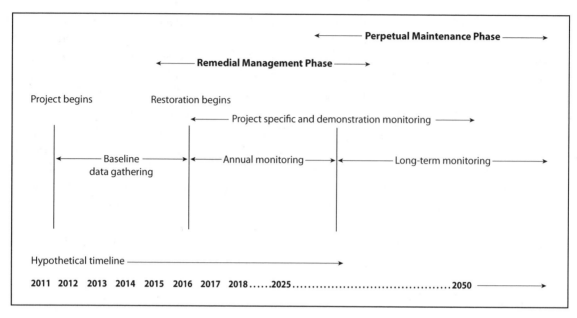

FIGURE 6.1. Monitoring for management and maintenance, hypothetical timelines

- Vegetation (vascular or nonvascular plant groups such as flowering plants, bryophytes, algae, etc.)
- Fauna (birds, small mammals, reptiles, amphibians, macroinvertebrates, fishes, etc.)
- Surface and shallow ground-water hydrology
- Water quality (chemistry)
- Woody and nonwoody biomass, including litter and standing-crop (living) biomass.
- Food, fiber, fuel, or mineral production or harvesting.
- Soil chemistry and organic matter.

Basic Approaches to Monitoring

Regardless of other monitoring used, qualitative monitoring is always used. In the least formal application, qualitative monitoring involves simple observation, something every farmer, forester, wildlife manager, or field biologist does continu-

ously. If you are going to restore an ecosystem, you will need to become a keen observer. The species that inhabit the ecosystems in your project are the most frequent focus of quantitative monitoring. Different organisms usually demand unique sampling approaches and represent different kinds of identification challenges.

1. *Observations.* The most superficial monitoring, and the most widely used, is simple observation. This is universally employed by those who manage or work with natural resources or ecological restoration. To be useful, it requires that you have good familiarity with the land; recognition of species, especially invasives; and a good understanding of the ecological processes you wish to restore. Get in the habit of carrying a notebook and camera, and diligently record your observations, questions, and insights.

2. *Repeated photography.* This method is a good augmentation of observations and provides a hard record in the form of photographs that can

be compared from one year to another. Use reference points that can be relocated. GPS is a great way to do this easily, but permanent, fire-proof markers, such as metal posts, can also be used. These should be given numbers and permanent GPS or map locations noted. Mark them on your basemap. Be sure to use comparable lighting, showing the scene toward the same azimuth (direction), at about the same time of year and time of day. Record dates and locations to permit comparisons from one year to the next. If possible, use digital images that can be easily associated with locations on your basemap.

3. *Aerial surveys.* Nearly all of the United States is flown by the government at least every ten years, often more frequently, with digital images available at cost. When more superficial monitoring is appropriate, such as with long-term monitoring and especially over large areas, aerial photography is a good way to assess changes in vegetation cover. With a bit of experience, even novices can learn to interpret aerial photographs sufficiently to monitor patches of invasive species, changes in tree cover, and comparable details at a scale of a few meters. See the discussion of aerial photographs in steps 1 and 2.

4. *Quantitative sampling.* Hard data are often needed for one or more of the reasons mentioned previously. Different kinds of organisms demand different quantitative sampling approaches (discussed later). Nearly always, however, vegetation will be included in monitoring surveys. As vegetation is being sampled, some faunal sampling may also be possible, or at least the same sampling design can be used to monitor some fauna.

Surveying Different Organisms

Plant surveys. The time required to do an adequate inventory depends on your familiarity with plants in your area. Even good botanists often carry bags to collect specimens of plants with which they are not familiar. These unknown plants can be the most interesting species, and should not be ignored. Unknown plants also can be photographed to create a permanent record. For instructions on preparing plant specimens, check with the local college to inquire about help in identification, or better, take a course in plant taxonomy. Most colleges that employ a botanist will have a herbarium where you can compare your specimens to those identified by experts. Field keys and manuals for plant identification are abundant. The best are those developed for your local area and habitat types. Check with the local library or college for suggestions, and browse local bookstores. Also, lists of species found in your area can probably be obtained from your state department of conservation or natural resources. You can often browse their website for such resources. At a minimum, get a list of common invasive species occurring in your region, and make a point of learning to recognize them.

Bird surveys. It is easy to learn to recognize birds by sight and sound. To capture the most useful information, surveys should be conducted early on quiet, sunny mornings during the peak of the breeding season. Minimally, you should compile a list of all species using each habitat type you have recognized in your restoration plan. Plan to spend five to ten minutes per acre in thick habitat, walking slowly, or better, standing or sitting quietly, watching and listening. Binoculars, a good bird guide, and your notebook are all that you need. There are many bird guides available, and each has some advantages and disadvantages, depending on your experience and location. We recommend doing an online search for "bird identification guides" and choosing one that seems best suited for your needs. Also check with a local bird club and your library for their recommendations. There are many very good amateur birders in nearly every community, and they often are willing to do bird walks with you, at least to get you started. There are dozens of sources of recorded bird songs, many with CDs that you can purchase. Look for those that are comprehensive for birds of your area. Public libraries sometimes have CDs that you can check out. Also, a free, online source is Patuxent Wildlife Research Center's checklists and bird songs (http.//www.mbr-pwrc.usgs.gov/id/songlist.html). WhatBird.com offers songs of 800 species of birds for an iPhone or iPod that can be easily used in the field. There are dozens of other online tools to aid identification by sight or sound.

Frog and toad surveys. If you have a stream, pond, or wetland on the property being restored, you might enjoy keeping track of the frog and toad species using them. This is most easily done during the spring mating period, when males are singing to attract mates. The Northern Prairie Wildlife Research Center has an online guide for amphibians in North America north of Mexico (http://www.npwrc.usgs.gov/resource/herps /amphibid/index.htm). We recommend a recently published book, *The Frogs and Toads of North America*,[1] which includes a CD of songs. It is not difficult to quickly learn those that occur in your region. Your state may have CDs and lists more specific to your area, so check with the local library or state natural resource agency.

Pitfall traps. A technique that can capture a variety of organisms that wander on the ground is pitfall trapping. We prefer to use 28 oz. (680 g) cleaned food cans with one lid removed. Punch small holes in the other end to allow water to drain in case of rain. Carefully remove only enough soil such that the upper edge of the can is flush with the ground, and only wide enough for the can. Scatter the soil away from the area, and disturb the vegetation and litter as little as possible. Leave cans for 24 hours and remove or cover them between sampling periods. Holes can be left for a few weeks, but may have to be re-excavated if sampling is done only once a year. Organisms collected can be placed in separate plastic containers and brought to a processing location where they can be identified and recorded by sample. Use as many traps as can be easily processed in a few hours, usually no more than a dozen or two, depending on experience with identification. Organisms can be preserved as reference specimens or returned to the field. The kinds of organisms commonly captured include wandering ground beetles, wolf spiders, salamanders, and ants, among various other invertebrates, depending on habitat. Again, check with the local library, college, and book stores, or check online for needed reference materials to aid your identification. Children enjoy helping with this kind of sampling, and it can be great family fun with many learning opportunities.

Snake boards. Many kinds of organisms, including snakes and scorpions, like to hide under structures. A sampling technique can take advantage of this behavior. Cut standard-sized pieces from old exterior plywood or similar material. A full sheet of plywood divided into two 4 × 4 foot pieces is a good size. Place these randomly or in a grid pattern in the area you wish to sample. Leave boards for at least a week or longer before checking them. Boards can be left for several years, and probably are more useful over time. Recognize, however, that vegetation under them will die, and if prescribed fire is used, they will burn.

When you survey boards, raise them carefully, especially in areas where venomous animals occur. Some insects and small mammals may scurry away before you can identify them, but try to identify each of the reptiles and amphibians that might be present. There is no reason to disturb the animals any more than necessary. Lower the boards carefully when you are done. You should number the boards and keep data separate for each. Some of the same individuals are likely to remain under the board and be observed on subsequent surveys.

Insect surveys. If you or your children have a good ear and enjoy listening to nature, you might want to extend your auditory survey to include insects. Crickets, katydids, and cicadas represent a relatively small percentage of insects in any terrestrial ecosystem, but they can be easily monitored during the breeding season because they make sounds that are identifiable. As with frogs and toads, you will need to get a good field manual with a CD.[2] Surveys can be either day or night for audible insects, and both are desirable. Butterflies can be monitored along the vegetation transects (described previously). Numbers of observations of species (frequency) should be noted. There are many excellent manuals for identification of butterflies and moths.[3] A free resource is "Butterflies and Moths of North America" found at www://butterfliesandmoths.org/faq/identify.

Aquatic invertebrates. Even if you have only an ephemeral or small stream in your project, improvement of water quality may be one of your goals. An easy, qualitative way to assess stream improvement is to monitor the kinds of invertebrates in the water. Those most useful for beginners are called macroinvertebrates because they are larger than the tiny organisms that, while very useful, require considerably more experience and training, and a microscope to identify. As with other kinds of organisms, there is a host of available resources. Choose one or more appropriate for your level of experience and region. A government-sponsored portal for exploring resources is the National Biological Information Infrastructure (NBII), at which you can find many resources useful in monitoring (http://www.nbii.gov). You can navigate from their homepage to "aquatic invertebrates," among dozens of other topics that may be of interest, including fire management. Another useful portal, more focused on aquatic invertebrates, is found at http://www.epa.gov/bioiweb1/html/invertebrate/html. Sample aquatic invertebrates in the late spring or early summer. To get useful information, use a standardized sampling approach described in many of the resources you can access online.

Quantitative Measurements

Well-stated goals often require quantitative data to evaluate. Many of the monitoring methods we describe can provide good quantitative data if sampling is done consistently and without bias. Here we discuss some of the most commonly used quantitative measures.

Plot size. The probability of encountering a species when sampling, called frequency, is proportional to the size of each sampling plot or the time spent making observations. Therefore, in developing a sampling plan, include the plot size or observation time to be used for each kind of organism and be consistent throughout the life of the project. For example, you might use a one square meter frame to estimate cover and frequency of herbaceous plants, but 50 or 100 square meter plots for shrubs and trees, or a five-minute sample for birds.

Small, circular sampling frames can be easily assembled using one-inch PVC pipe available at any hardware store. Carefully calculate the circumference of the circle to provide whatever size plot area you wish, called a quadrat. For example, a one square meter plot has a circumference of 3.545 meters (11 feet, 7.6 inches). This should be the inside measurement of the pipe when pulled into a circle. This is easily accomplished with a plug tapered at both ends such that the open ends of the PVC pipe can be pulled together. Plots of hundreds of square feet or larger are usually square or rectangular. They can be laid out using a measuring tape and temporarily marked with flag stakes and string.

Sampling scheme. To eliminate bias, sampling is done using a systematic, random, or combination scheme (see figs. 6.2a and 6.2b). Select a method that can be repeated year after year. This involves permanent marking of baselines and transects for vegetation and birds, and good records on date of sampling and size of plot used.

For each area to be sampled, locate a baseline either through the habitat or along an edge. Along the baseline, mark points for starting transects systematically (at regular intervals) or randomly (distances determined by random numbers table) using permanent markers such as metal stakes. Extend transects at right angles from the baseline at each point using a compass azimuth (direction). Locate plots either randomly or systematically along each transect, noting distance and direction from baseline. Leave a buffer around the area to be sampled to avoid edge effects. For birds, a buffer of 100 or more feet is desirable, whereas for invertebrates, small

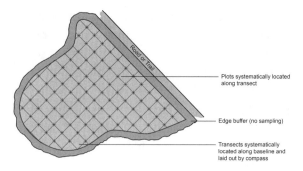

FIGURE 6.2A. A systematic location of transects from baseline

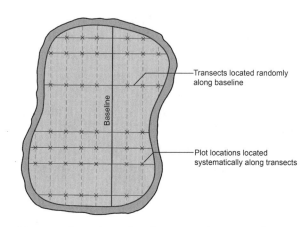

FIGURE 6.2B. A random location of transects from baseline

mammals, or vegetation, a buffer of 20 to 30 feet should be sufficient. The number of plots depends on variance and desired accuracy of your estimate. More is often better. Forms for recording herbaceous vegetation cover data (vegetation data form 6.1) and woody vegetation cover (vegetation data form 6.2) are provided in appendix 1.

Commonly Used Measures

There are many ways to summarize quantitative data, some good and some not so useful.

Cover of vegetation is one of the most widely used metrics of a terrestrial community. It is estimated in conjunction with frequency. Because it is impractical to sample all vegetation in even a relatively small restoration project, subsamples are taken in some systematic or random way (sampling techniques are discussed later). Average values provide good estimates of abundance, and the frequency with which a species was recorded is a measure of distribution. Cover of species is easily estimated visually, and is correlated to biomass. Use a small quadrat of a few square feet or less for vegetation that is less than a meter tall, enabling you to look down on the vegetation and assess the cover of each species as a percent of the quadrat. Place the quadrat carefully into the vegetation, and estimate the percent cover for each species. It is impractical to re-locate small plots precisely, but sampling along permanently marked transects at regular intervals is possible.

We have found that a line-intercept method is much easier for estimating cover of trees and shrubs. Using a standard length of a transect, say 50 meters, note the beginning and ending point for the crown (canopy) of each species along the transect (see fig. 6.3). Add the intercepts for each species along each sample transect, and divide by the total length of the transect to obtain a percent cover for the species. (See woody vegetation cover data form 6.2, appendix 1.)

Richness and diversity. A listing of species present provides the simplest measure of composition. The number of species found, called richness, is one expression of diversity. There are

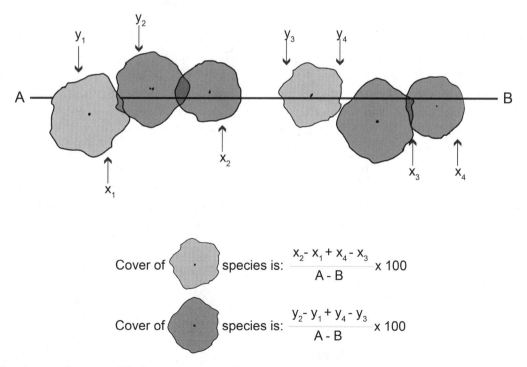

$$\text{Cover of} \quad \text{species is:} \quad \frac{x_2 - x_1 + x_4 - x_3}{A - B} \times 100$$

$$\text{Cover of} \quad \text{species is:} \quad \frac{y_2 - y_1 + y_4 - y_3}{A - B} \times 100$$

FIGURE 6.3. A schematic of the line intercept method

many ways to estimate and express diversity, some better than others. Regardless of which method is used, keep in mind two points: (1) diversity in one group of organisms, plants for example, generally parallels diversity in others, insects or birds, for example; (2) only a subset of the total diversity of any ecosystem can be estimated. We recommend monitoring diversity in vegetation and at least one consumer group, for example, birds or insects. If your project includes a stream or wetland, consider monitoring amphibians and macroinvertebrates.

The minimum level of biological monitoring involves lists of species. Such lists, however, are meaningful only if a systematic and standardized search is used, and if the observer(s) has reasonable competency in identifying species or taxa, or at least the ability to recognize differences among them. We use *taxa* to indicate related groups of species, usually genera or families. Especially with invertebrates, where identification to species, and sometimes to genera or families, requires specialized knowledge, identification might be less specific. Some species of plants are often grouped as well, sedges (*Carex* spp.) for example. Birds, amphibians, and most vascular plants, on the other hand, can be recognized to species with a bit of practice, and assistance when needed. With few exceptions, all biological monitoring, including quantitative methods discussed later, depends on recognition of species or taxonomic groups. Especially with vegetation, it also is useful to categorize taxa as native desirable (ND), native invasive (NI), and nonnative (NN). Increasing numbers of ND and decreasing numbers of NI and NN species indicate that the system is moving in a positive direction.

Frequency of occurrence and abundance. Measures of plant or community structure usually in-

volve estimates of plant abundance, including cover, broken down by type of plant. For example, sampling can be designed to estimate cover of trees and snags; understory small trees, vines and shrubs; ground layer (often defined as plants less than one meter tall); bryophytes and lichens; coarse woody debris; and fine litter. Different techniques or plot sizes are usually required for different types of plants. Often numbers of stems per unit area, or density, is also estimated, especially for woody vegetation. Sometimes biomass of vegetation is used as a measure of abundance, although that requires that plants be harvested. With a systematic or random sampling design, *frequency* of occurrence of taxa can be determined. Frequency is the percent of samples in which a taxon occurs, a function of distribution and abundance. High frequency means that a taxon is widespread and common, whereas low frequency means it is local or uncommon.

If vegetation is not sampled, it still is useful to add an abundance index to species lists to improve the quality of the survey: widespread and common (C); irregular occurrence (IO); local, found only in one location but with many individuals present (L); and rare, only one or, at most, a few individuals found (R).

Conservatism indexes. Some ecologists have developed conservatism indexes to augment plant data. Some species are found only in very high-quality habitats while others, including what are commonly called weeds, are more ubiquitous. Each species is subjectively assigned a conservatism value (C value) depending on the degree to which the species is thought to be restricted to relatively undisturbed, high-quality habitats. This technique is called floristic quality indexing (FQI). A study recently demonstrated, however, that species richness measures derived from a summation of the number of species per plat (data forms 6.1, 6.2, appendix 1) better reflected quality of restored prairies.[4] The inclusion of species with high C values in the restoration apparently led to artificially high FQI values, whereas actual richness numbers better reflected the overall quality and changes of the plant community.

Timed meander searches. Species richness and diversity of all kinds of organisms can be sampled with a timed meander search (TMS) technique. For plants or birds, TMS involves slowly walking though each community and listing new species recorded in succeeding increments of time. It is most easily done with two people, one to keep track of time and one to list species as they are found. After each minute, a line is drawn under the last species added to the list, and the process continues until no new species are found in several minutes of searching. One can obtain similar information by plotting species added as a func-

tion of the number of plots or samples surveyed. The data contribute to the development of total species lists and help quantify diversity for each community. The more diverse the community, the longer the time (or the more samples) required for the number of species plotted over time to level off (see fig. 6.4). The same technique can be used to compare diversity of macroinvertebrates in aquatic environments. With macroinvertebrates, either search time or number of samples is plotted along the horizontal axis.

Importance values. Additional information can be obtained by expressing the relative frequency and relative cover of each species of plant in the sample collected and recorded in data form 6.1 and 6.2 (appendix 1). For plants, one might use relative frequency and relative cover numbers. Relative values (see table 6.1) are calculated by adding the total frequency and cover values for all species, and dividing the sum into

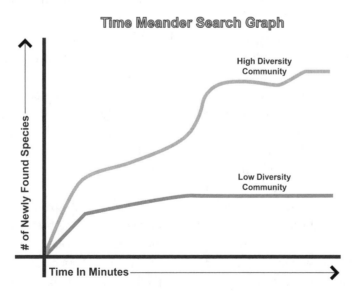

FIGURE 6.4. Time meander search graph

TABLE 6.1.

Hypothetical computations for vegetation cover, frequency, and value

	Avg. cover	Rel. cover	Obs. frequency	Rel. frequency	Importance value	Rel. importance value
Species A	72.5	32.8	65	37.4	70.2	35.3
Species B	51.9	23.5	48	27.6	51.1	25.7
Species C	32	14.5	21	12.1	26.6	13.4
Species D	63.7	28.8	38	21.9	50.7	25.5
Species E	1.9	0.1	2	0.1	0.1	0.1
Total	222.0	99.7	174.0	99.1	198.7	100.0

the average frequency and cover (or density) for each species. This relative value is expressed as a whole number between 0 and 100. The sum of the relative frequency (maximum of 100) and relative cover (maximum of 100) for each species is a value between 0 and 200, 0 if the species was not recorded and 200 if only that species was present. This provides an index of how important each species is relative to others in the sample. Decreasing importance of NN and NI species and increasing importance of ND species over time is a good indication that the restoration is moving in the right direction.

Point counts. Birds can also be quantified. Establish random or systematic points. These might be a subset of the same points used for monitoring vegetation. Points used for sampling birds should be a minimum of 300 feet apart. Sample as many points as time permits, but avoid any within 100 feet of the edge of the area being sampled. During the peak of the breeding season, on a quiet, sunny morning, beginning at first light, visit each point. Use a watch to control for time, and spend exactly the same amount of time at each point. Five minutes is adequate. Using a compass to orient the plot and the bird point count data form provided (data form 6.3, appen-

dix 1), record the location of each bird heard or seen within 150 feet during the five-minute period. Measure or pace 150 feet until you can estimate this distance consistently with reasonable accuracy (+/– 10 feet). Try to avoid recording the same bird more than once. Each point in the surveyed area represents one sample. From these data, you can compile a list of species and calculate density (number of species per unit area) and frequency (percent of points at which a species was recorded) for each species.

Bird data is made more valuable when it can be determined whether the species observed are breeding in the habitat. Most states now have a breeding bird atlas program in which experienced observers routinely survey the state for breeding birds. Evidence that a species is breeding in a restored ecosystem is a much stronger basis for evaluating habitat than mere presence. Widely used criteria for breeding status are the following:

1. *Observed*: A species, male or female, was observed during the breeding season, but no evidence exists to indicate the species is breeding.

2. *Possible*: A species, male or female, was observed in suitable habitat and at a time during the breeding season that indicated it was possible that

breeding occurred. Singing males often indicate possible breeding.

3. *Probable*: Several types of observations indicate that a species is probably breeding: a pair (male and female) observed in suitable habitat, a permanent territory as indicated by multiple observations of a singing male, male-to-male conflicts, courtship or copulation, or agitated behavior when people are present.

4. *Confirmed*: This is the most restricted indication of breeding. Observations to confirm breeding must provide direct evidence that the species is breeding at the site. Nest-building by species other than wrens, thrushes, or woodpeckers; distraction displays; a nest with eggs; a used nest with eggshells; adults carrying a fecal sac or food; recently fledged young; or a nest with young seen or heard.

Physical and Chemical Monitoring

For physical and chemical environmental monitoring, such as soil organic matter and pH, or hydrology, long-term quantitative methods with systematic or random sampling are generally required. The time and cost of doing quantitative physical and chemical monitoring preclude its use except where needed to detect subtle changes or for examining responses to treatments. While ideal, and certainly something to be carefully considered, the justification for the added time and cost will depend on the goals and objectives of the landowner/manager. Unless your monitoring is contracted, you usually will gather samples to be sent to state or commercial laboratories for analysis, as discussed later.

Soil organic matter. O.M. content in soil is spatially variable, but tends to increase as ecosystems are restored to full health. Dry samples are weighed and subjected to either high temperature in a furnace or wet oxidation using a strong basic solution. Loss of mass results from oxidation of organic compounds that are released into the air as carbon dioxide. Contact your local NRCS or extension office for costs and procedure for mailing or dropping off samples to be tested at a state-certified lab.

Collecting soil samples for analysis should be done systematically. An inexpensive soil probe is ideal, but a shovel also can be used. Avoid sampling when the soil is wet if possible. We recommend a systematic sampling scheme to avoid bias. Follow the instructions that come with the testing kit.

pH and other chemical tests. The same soil sample being tested for O.M. can be tested for other variables that may be of interest. pH, or soil acidity, is often important. Depending on site history, especially past agricultural practices, fire, and hydrologic changes, pH can vary considerably. As pH changes, availability of many nutrients and activity of soil organisms change. pH may require adjustment before introduction of some native species. Testing soil reaction is a simple and inexpensive test. A LaMotte soil pH kit can be purchased from any forestry supply company for about $75. Used carefully, this kit is sufficiently accurate for restoration-related needs. Other chemical tests may include levels of potassium or phosphorous, or tests for potentially limiting toxic contaminants. Phosphorous often accumulates in soils, especially those that are alkaline and have been farmed for many years. These tests must be done in a laboratory with proper equipment and control. Costs will vary, but all chemical testing can be performed using the same sample as collected for O.M.

Water testing. If you have a stream within your project, direct monitoring of water quality is an option. There are many parameters to consider in addition to macroinvertebrates (discussed previously). The most important measurements usually are dissolved oxygen and nitrates. Unless you have a commercial or state-run laboratory nearby, you may need to invest in a Hatch Kit at a cost of about $325 (available from forestry supply companies). This kit has the tools and reagents needed to run a wide range of tests, including dissolved oxygen and nitrates, phosphate, and ammonia, among others. Perhaps you can interest a local school in using your stream as a laboratory. Many have the equipment needed and often are interested in working on local projects.

Hydrology. Although hydrology is extremely important, monitoring change is very challenging because of variation in the landscape and climate. In many projects, restoration of hydrology is essential, and in nearly all projects there will be changes associated with alteration of stream hydrology, removal of drainage tiles and ditches that were installed to dewater areas, or decreasing runoff and increasing infiltration as native vegetation is restored. There are a few relatively simple ways to monitor hydrology, but in all cases, many samples and long-term data are required to establish trends or change.

Low-tech, inexpensive wells can be used to monitor changes in ground water. Hand auger a 1- or 2-inch hole four to six feet deep and immediately insert a 1- or 2-inch diameter PVC pipe. The pipe should have several small holes drilled in the bottom six inches to admit water, and a cap glued over the end to close it. Attach a cap to the top that can be easily removed. These tubes are left in place between sampling. Use a small,

retractable-metal measuring tape, a string, and a fluorescent yellow or orange *water-soluble*, Magic Marker pen. Before each measurement, color enough of the string to reach the length of pipe. After allowing the ink on the string to dry for a minute, carefully insert the zero end of the tape down the well to the bottom with the string attached. After 10 to15 seconds, *slowly* pull up the measuring tape and locate on the string where the water-soluble ink was dissolved. This depth on the tape identifies the depth to the water. By subtraction, you can compute the distance from the top of the ground to the water surface. Groundwater will fluctuate over the year, so repeated measurements will be necessary to establish trends. Several monitoring wells can be installed along a hydrologic gradient, for example, from a wetland up a slope to higher ground, to gain a better idea of changes in ground water.

Soil moisture is extremely variable, making it difficult to establish change. A variety of commercial probes is available that give reasonable approximation of soil moisture, but they are expensive, ranging from $300 up. Very accurate methods require calibration, usually over two or more wetting and drying cycles. The least costly way of measuring soil moisture is a gravimetric method based on accurately weighing fresh samples, drying them at 100°C–110°C (212°F–230°F), and reweighing them. Drying should continue until the samples reach constant weight. This can be done in a kitchen oven with the door ajar. The loss of weight, expressed as a percent, is the amount of water the sample contained.

Changes in stream flows are even more variable than soil moisture, but if a project is conveniently located near your home, you might wish to establish long-term monitoring on it. Choose a place along the stream that is contained by a permanent structure such as a bridge abutment or bedrock. Or if the bank and bed are very stable, you probably can use a point without hard structure to get reasonable results, but recognize that any change in the reference point will require that you recalibrate discharge. Stretch a level string across the stream, and drop a tape measure from the string to the bottom at frequent intervals from one side to the other. Plot these measures

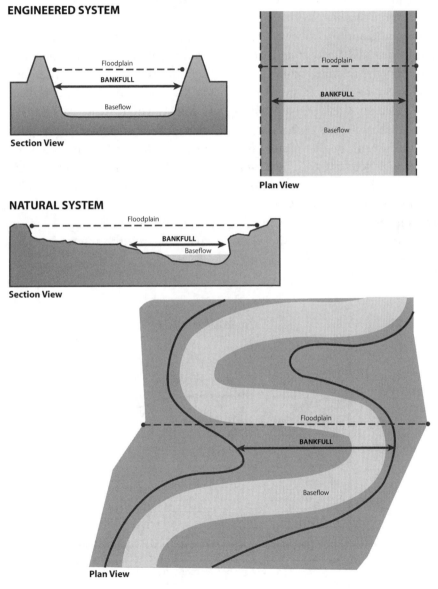

FIGURE 6.5. Stream channel characteristics for cross section

on a piece of graph paper to develop a cross-section of the stream. Add a scale to the diagram so that you can calculate the number of square feet below any level (see fig. 6.5). As stream level changes, estimate velocity through this reference area by using a floating object tossed into the stream a few yards upstream (use a tape to measure the distance) and a stopwatch. Repeat the observation several times and use the average in feet per second for the time it takes the floating object to cross through the reference point. Multiply the velocity times cross-sectional area to obtain cubic feet per second. After you have determined the volume discharge for a wide range of staging levels of the stream, graph volume as a function of stage height (see fig. 6.6). Thereafter, you need only read height at the reference point to determine volume from the graph. Once you have these basic stream measurements you can prepare some secondary calculations using data form 1.10 (appendix 1). As with soil moisture,

there is a wide variety of high-tech devices for monitoring stream discharge, but cost is prohibitive unless you are doing a lot of stream-gauging work.

Monitoring Equipment

Equipment used in environmental monitoring varies from none to extremely expensive devices. Simple, less expensive monitoring is generally sufficient for small, simple projects, or even large projects where greater accuracy and detail is unnecessary. For large, complex projects or projects where extensive monitoring is required, hiring a consultant to do the monitoring is generally advisable. The equipment required for most projects not overseen by professionals can be found by scanning the catalog of a forestry supply company. Search the internet to locate one of your choosing.

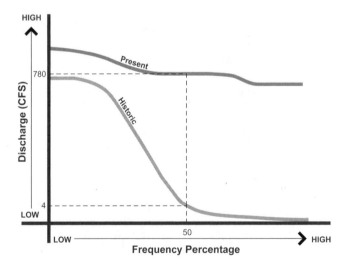

FIGURE 6.6. Example stage-discharge graph for representative stream

Notes

1. Lang Elliot, Carl Gerhardt, and Carlos Davidson, *The Frogs and Toads of North America.* (New York: Houghton Mifflin Harcourt, 2009). Book and CD.

2. Lang Elliot, and Wil Harshberger, *The Songs of Insects.* (New York: Houghton Mifflin, 2007). Book and CD.

3. Robert M. Pyle, *The National Audubon Society Field Guide to North American Butterflies.* (New York: National Audubon Society, 1981).

4. Marlin Bowles, and Michael Joes, "Testing the Efficacy of Species Richness and Floristic Quality Assessment of Quality, Temporal Change, and Fire Effects in Tallgrass Prairie Natural Areas." *Natural Areas Journal* 26(1): 17–30.

Implement the Plan

The beginning is the most important part of the work.

Plato

A few additional things need to be done before actually starting work on the land. This includes a review of any regulatory requirements that, if ignored, could result in costly delays and penalties. You also need to develop a phasing plan for the work, then reconsider and refine your budget with estimates for each phase (see fig. 7.1). A phasing plan divides the project into subprojects that efficiently structure the implementation. Finalizing plans may also involve procuring plants and seed or other supplies and equipment. These should be grouped to receive the best prices, and according to when they are needed in the project. Once this is done, you can prepare your implementation schedule for the first five years, during what we have called the *remedial restoration phase*.

Task 21. Develop a Task List and Schedule

To transition from planning to implementation, a schedule of projects and activities is needed. To illustrate, use the restoration scheduling data form (data form 7.1, appendix 1) for a project. To clarify where we are in the overall restoration process, the additional restoration steps, including some that precede actual work on the ground, are listed, which you should add to the data form.

1. Completion of a five-year remedial restoration plan
   Develop project phasing schedule
   Permitting
2. Procurement of seeds, plants, and other needed materials (addressed in step 9)
   Purchasing
   Contract growing
   On-site growing and harvesting
3. Contracts for outsourcing work (addressed in step 9)
   Procurement of contractors and materials
   Budget refinement by phase and task
4. Financing restoration projects
   Finalize financing plan
   Obtain project financing
5. Commence remedial restoration
6. Review and refine remedial restoration (addressed in step 9)
   Monitor and review success, costs, and process
   Adaptively refine methods and budgets

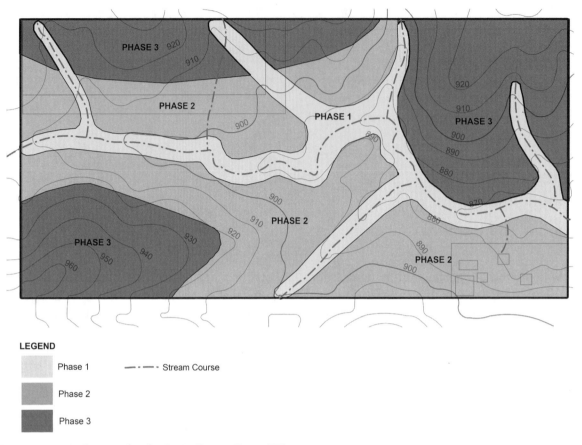

FIGURE 7.1. A phasing plan for Stone Prairie Farm, WI

7. Begin maintenance restoration
   Finalize maintenance restoration plan
   Initiate maintenance restoration
   Monitor and review successes and failures
   Adaptively refine methods and budgets
8. Complete implementation
   Keep meticulous records (step 9)
   Share your experiences (step 10)

If you lack the time, adequate labor, or equipment to do some or all of the tasks, you need to begin the process of obtaining bids for services and materials well ahead of implementation. For larger projects, you will need to negotiate various agreements or contracts for services or materials. After you obtain the bids, you may need to revise project budgets, perhaps even implementation plans. Later, we describe a series of strategic approaches to obtain services and materials that can save you money and time, increase certainty in financial planning, and ensure that the work gets done as scheduled.

Once you have financing and finalized plans, and have lined up needed services and materials, you are ready to begin work. Make sure as you arrive at this point that you still are on schedule to implement each task at the proper time. As previously discussed, many tasks such as seeding and planting or applying herbicides are time sensitive. If for any reason you have missed the seasonal window, revise your schedule or consider alternative plans for reaching the goal. During both the remedial restoration phase and the maintenance restoration phase, the process of restoration involves (1) executing the planned work, (2) recording the accomplishments, (3) monitoring the responses on the land, (4) checking and revising budgets and costs, and then (5) refining the overall restoration program to adaptively improve and refine the project.

It is during implementation of restoration programs that you can actually evaluate the efficacy of your plan. If the plan was well developed, you will have plenty of successes but nearly always some setbacks and unexpected challenges. This is to be expected and does not necessary imply a poorly developed plan. Take the monitoring and adaptive refinement process very seriously, as it can save time, money and, most important, it can accelerate effective and successful results over the course of your project.

Finalization of Plans

Determining where to begin and how to proceed across a landscape confounds many restoration projects. We have learned that there are strategic considerations that can guide you in these decisions. For example, if you have a high-quality remnant area, often a good source of seed for expanding the restoration efforts over the larger landscape, make it a high priority to mitigate stressors that are affecting it. Likewise, if you have degraded patches of an ecosystem that are likely to recover quickly with minimum effort, include these in your earliest efforts. Beginning restoration on larger disturbed areas such as conversions

from row-crop agriculture are generally guided by practical issues of cost and labor.

Task 22. Develop Project Phasing Plans

Phasing plans are typically prepared for restoration contractors prior to the bidding process. This information informs the person doing the work how many and what kind of resources are needed at each step. With each task, prepare a schedule for initiation. Because many tasks depend on completion of preceding tasks, you also will need to estimate time to complete for many. There are usually three fundamental drivers that influence the phasing of work during a restoration project:

- *Ecological:* These plans are designed using natural breaks in the land to define work phases and boundaries. Defining restoration units should be designed to protect and enhance ecotones (discussed in step 6) and their management needs. With units defined, the timing and sequence of work will depend on seasonal conditions for optimizing success, and the priorities established in your plan. Also keep in mind that there is often a logical sequence to activities on the land. For example, site preparation precedes seeding or planting, and both may have seasonal windows for completion.
- *Costs:* What you can afford to do at any time is the basis for both a schedule and a phase plan. Where money drives the availability of resources to implement a plan, you can start the phasing layout with some annual maximum or average expenditure in mind, and then adjust plans annually if things cost more or less money than expected.

- *Labor*: Having adequate labor available at critical times is commonly a limitation for initiating work. It is not uncommon for a contractor to lose some staff, or they may be overcommitted. The planting or herbicide application window is often quite narrow. Contractors typically pay attention to their most profitable projects. Contracts with penalty clauses for noncompliance also take priority. Especially if your project is small and you have a lenient contract, be prepared for setbacks. Even with a good contract and a reliable contractor, weather conditions can trump your plans. Design your phasing plan with as much flexibility as possible. For example, if you miss a window on one unit, reschedule it for the next window, and focus on units where work can be done.

Financing the Restoration

Except for minor work, such as a small prescribed burn, chances are the costs may exceed the funding available at the time all desired work could be completed. Coming up with the cash to cover the costs of the restoration can require some creative thinking. At Stone Prairie Farm, Steve came up with the following solutions:

- *Renting land to a neighboring farmer*: Rent portions of the land to a farmer for soybean production for a year or two. Cultivation of soybeans eliminates invasive grasses and other weeds, and provides some income for other projects in the meantime. Steve also required the farmer to remove all interior fences as a part of the rental agreement.

Normally, farmers do weed control in their crop fields, so this represents no additional cost. However, the fence removal did add to the farmer's costs, so Steve negotiated a reduced rent by 20 percent.

- *Growing cover crops for sale*: If soybeans are grown, in many areas winter wheat, or some other suitable small grain can be planted in the soybean stubble. Steve planted winter wheat and immediately after the wheat was harvested, a barley crop was planted. The sale of grain provided revenue for purchase of native seed and plants.

- *Crop subsidies*: Many farmers receive crop subsidies from the US Department of Agriculture (USDA). Consider phasing out production as you implement the restoration. By continuing subsidies until a unit is ready to be restored, you maintain some income that might be applied toward restoration costs. Be aware that removing acreage from agricultural production can result in an increase in real estate taxes. Farms often enjoy a reduced tax rate, and the increase in taxes can be considerable. Check with your county auditor.

- *Other federal assistance programs*: Steve used federal assistance to address a seriously eroded stretch of stream bank. He asked the NRCS to design and cost-share (75 percent) the repair and stabilization of a stream bank damaged by cattle. Steve describes the "baggage" that came with this funding source in his book, *Nature's Second Chance*, and also how he worked with NRCS to develop more environmentally friendly plans.

- *Harvesting seed from previous plantings to seed later projects*: As units were successfully established, Steve was able to both sell seed to a prairie nursery as well as harvest it for his own use. He also bartered with the nursery to provide seeds and plants in exchange for their harvesting from his early restorations. This allowed him to establish some rare species that he otherwise could not have afforded early in the restoration.

- *Bartering for labor*: An annual event at Stone Prairie Farm is the prescribed burn of the prairies, wetlands, and savannas. An arrangement was developed with Applied Ecological Services (AES), who wanted to practice and train new staff in prescribed burning. On an afternoon and evening when the conditions are right for conducting a prescribed burn, a group assembles and conducts the burn with oversight and instruction from the experienced staff. Afterward, they all share a kettle of spaghetti and a case of beer while debriefing the event. There are many organizations interested in learning and practicing prescribed burning, or other projects, such as identification of nonnative species. Look for opportunities to engage others in your project.

There are many other sources of funds to help finance restorations:

- *Home or farm equity loans*: If you own real estate with good equity you may be able to obtain a loan with relatively low interest, especially if the money is to be used in conjunction with the real estate. This is commonly done to repair or upgrade buildings and facilities, so it is not a big stretch to think in terms of upgrading the land.
- *Partnerships and cost sharing*: If you have neighbors with similar interests, or friends who share your interests, some creative partnering is possible. For example, one colleague had urban friends help finance the installation of a fruit orchard. Another land owner got financial assistance from friends in exchange for hunting privileges. A farmer permitted friends to cut firewood in exchange for implementing his forest restoration plan. The opportunities for partnering are endless and depend entirely on your creativity.
- *Grants and soft money*: Grants are available in many regions for doing restoration, teaching others to do restoration or land management, or for solving real problems on your land that can be replicated in other properties with similar problems. If you have sufficient experience, you may be able to secure some funds and labor by training others, such as teaching restoration or management classes through a local vocational college or high school, using your land as the hands-on laboratory. Even short courses such as workshops sponsored by various agencies or organizations can be a productive partnering strategy. There have been funds available for teaching stream and watershed restoration if the training can be related to downstream water quality and flood relief. The US Enivronmental Protection Agency (USEPA), NRCS, Corps of Engineers, water management agencies and drainage districts, and schools have access to these types of funds, or you can offer to work with these agencies and organizations to develop proposals. It often is also possible to develop partnerships with commercial establishments such as native plant nurseries. They like this exposure and the

opportunity to train new staff, and it never hurts to become known as a local expert in the eyes of the regulators and land management agencies.

Another source of money that is slowly becoming available involves payments for ecosystem services. If improvements in your land help address such problems as flooding or water quality, then you might be eligible for compensation for the benefit provided. Other opportunities might also be possible, such as wetland mitigation. AES (see www.appliedeco.com) and partners have become national leaders in developing wetland banks to mitigate wetland lost to development, and restoration to mitigate flooding and water quality problems. If you want to explore these kinds of opportunities, you will need to work with experienced contractors. Your local NRCS office or county conservationist might be able to advise you as to whether there are opportunities within your project.

A related source of funding may be through carbon sequestration in the near future. Grasslands, wetlands, and forests that are being restored represent long-term carbon storage through increases in soil organic matter. Carbon storage, of course, helps mitigate climate change.

Compensation for ecosystem services is still largely in the development phase. Many questions and issues are yet to be resolved, including regulation and monitoring to evaluate benefits being claimed. The best way for private landowners to tap these opportunities is through partnering with companies or organizations that have a demonstrated track record. There are thousands of individuals and companies trying to become established in this arena, and agencies are struggling to deal with the challenges of manag-

ing these opportunities. Be cautious and ask lots of questions.

Steve is exploring the development of *ecological reserves* (see *Nature's Second Chance*), in which multiple revenue streams and partnerships are developed to not only address restoration but to expand projects to a larger landscape by involving neighboring landowners. This is the ideal approach to restoration, given the way most property boundaries cut across watersheds and landscapes. This collaborative approach can lead to new sources of funding, sharing of equipment and labor, and opportunities to address some stressors that cannot be addressed in only portions of watersheds. This approach can link small parcels to take restoration to a meaningful scale, with an expanded range of associated ecological services and benefits, increasing financing opportunities.

Task 23. Develop Detailed Specifications

If you choose to work with contractors, you need some standard documents for explaining your project and formalizing relationships. We have chosen several examples and boilerplate contract types that you may find useful (see appendix 3).

A *scope of work* defines the job you want done, including what, where, when, and how. Obviously, we can only provide examples of how such a document is structured, which is useful to illustrate the expected and necessary content. In the appendix, we have provided a detailed document including site preparation, planting, and establishment period management and maintenance. A separate document is provided for scoping a simple prescribed burning project. We also have included an example of a purchase agreement for plants and seed materials that you may find useful. On the book website (www.restoringecologicalhealthtoyourland), we offer other examples and templates that may be useful.

Task 24. Determine Restoration Methods and Equipment Needs

There are nearly endless types of equipment options for addressing various restoration tasks; our aim here is to help you evaluate the purchase and operational costs. You will need to decide whether to purchase or rent equipment, or to hire the work done by a contractor. In considering whether to buy, rent, or hire, do not forget to consider the money and time required to learn to safely operate and maintain the equipment. If you own farming equipment, you know the maintenance and operational costs, and you can certainly use what you already have. Most restoration equipment is used intermittently and sits idle most of the time. Contractors, on the other hand, can amortize their investments over many projects and much greater use. Unless you have a huge project, or many projects, it usually is cheaper to partner or barter for the work you need done, or to hire a contractor, than purchase and maintain major items of equipment.

An equipment costing and decision matrix for your use in evaluating equipment investments (see table 7.2) allows you to weigh the alternatives of purchase or lease, bartering, or hiring contractors. The matrix does not include some of the most specialized equipment, such as specialty drills that are necessary for sowing many native plant seeds, nor equipment that requires specialized training, such as for prescribed burning. It is not possible in this matrix to reveal with

certainty hidden costs, such as delivery costs for new equipment, or training costs. For example, where the ecosystem you are restoring has fire requirements, if you intend to do the burning yourself, you will need training in the use of prescribed fire.

Supplies and Contractors

Lining up supplies and contractors can be time consuming and frustrating. Sources are not so difficult to find, especially when searching online, but getting the right material or service, when you need it, from a reliable source is the challenge. We focus here on some of the pitfalls and provide guidance to facilitate your search and decisions.

Task 25. Procure Plants, Seeds, and Materials

The final tasks in restoration planning are finalization of the plant, seed, and material lists, the phasing plans, and final revision of the project budget. The list of plants and seeds desired is best based on the reference natural area species data (see step 2). Most native plant nurseries will work with you to refine the lists. They can assist with the quantities and costs for seeds and plants, and the expected availability of the materials when you plan to begin your work. AES, for example, has completely computerized databases that are updated regularly with the supplies, availability, and pricing for hundreds of native plant nurseries nationwide, and it routinely provides assistance to restoration project developers, other restoration contractors, and landowners to prepare

TABLE 7.1.

Equipment cost and decision matrix

Machine	New cost	Service life (hours)	Annual use (hours)	Annual fixed cost	Fixed costs hour	% new cost	Repairs costs hour	Variable costs hour	Total machine hours/acre	Labor hours/acre
Plow 1-14″ MTD (mounted)	$215	1000	100	$40	$0.40	80	$0.17	$0.57	2.40	3.00
Plow 2-14″ MTD	$1,100	1000	100	$188	$1.88	80	$0.88	$2.76	1.20	1.50
Plow 3-14″ MTD	$1,600	1000	100	$274	$2.74	80	$1.28	$4.02	0.80	1.00
Tandem disk 6′	$1,250	1200	120	$219	$1.83	65	$0.68	$2.51	0.34	0.43
Tandem disk 8′ MTD	$2,800	1200	120	$478	$3.98	65	$1.52	$5.50	0.27	0.34
Rotary tiller 8-hp reartine	$1,150	2000	400	$308	$0.77	65	$0.37	$1.14	4.00	5.00
Rotary tiller 38″ MTD	$745	900	90	$127	$1.41	65	$0.54	$1.95	1.30	1.63
Rotary tiller 72″ MTD	$1,490	900	90	$251	$2.79	65	$1.08	$3.87	0.65	0.81
Cultimulcher 10′	$3,550	700	70	$618	$8.83	65	$3.30	$12.13	0.21	0.26
Harrow 10′	$266	1200	120	$47	$0.39	65	$0.14	$0.53	0.12	0.15
Plastic layer	$400	1000	100	$81	$0.81	70	$0.28	$1.09	2.00	2.50
Fertilizer spreader spinner	$395	350	70	$122	$1.74	75	$0.85	$2.59	0.12	0.15
Fertilizer spreader 8′	$900	350	70	$264	$3.77	75	$1.93	$5.70	0.42	0.53
Bed shaper	$590	500	50	$112	$2.24	70	$0.59	$2.83	2.00	2.50
Sidedresser 2-row	$860	900	90	$151	$1.68	65	$0.62	$2.30	0.55	0.69
Transplanter 1-row	$475	600	60	$89	$1.68	75	$0.59	$2.07	3.33	4.16
Transplanter 2-row	$965	600	60	$175	$2.92	75	$1.21	$4.13	1.67	2.09
Potato planter 2-row	$3,250	600	60	$554	$9.23	50	$1.63	$10.86	1.67	2.09
Plastic transplanter	$640	600	60	$113	$1.88	75	$0.80	$2.68	3.33	4.16
Planter 1-row	$400	1200	120	$70	$0.58	75	$0.25	$0.83	1.79	1.79
Planter 2-row	$1,050	700	70	$181	$2.59	75	$1.13	$3.72	0.89	0.89
Multivator 2-row	$2,790	825	165	$742	$4.50	85	$2.82	$7.32	0.75	0.75

Cultivator 1-row	$475	840	120	$106	$0.88	85	$0.48	$1.36	1.38	1.36
Cultivator 2-row	$618	840	120	$140	$1.17	85	$0.63	$1.80	0.69	0.69
Sprayer w/boom or gun	$695	750	150	$192	$1.28	80	$0.74	$2.02	0.21	0.21
Sprayer drop nozzles	$750	750	150	$207	$1.38	80	$0.80	$2.18	0.43	0.43
Air blast sprayer	$4,600	1000	200	$1,252	$6.26	80	$3.68	$9.94	0.63	0.63
Mower 7'	$2,012	1000	100	$340	$3.40	180	$3.62	$7.02	0.66	0.66
Rotary mower 5'	$918	1000	100	$157	$1.57	180	$1.65	$3.22	0.89	0.71
Rotary mower 7'	$1,550	1000	100	$263	$2.63	180	$2.79	$5.44	0.75	0.60
Potato digger 1-row	$3,300	1000	100	$551	$5.51	50	$1.65	$7.16	4.13	3.30
Front-end loader	$2,650	2500	250	$445	$1.78	65	$0.69	$2.47	0.00	0.00
Forklift 1-ton	$11,500	6000	600	$1,926	$3.21	90	$1.73	$4.94	0.00	0.00
Hole digger 24"	$800	1000	100	$136	$1.36	60	$0.48	$1.84	0.00	0.00
Harvesting aid 2- to 3-row	$3,450	2000	200	$634	$3.17	50	$0.86	$4.03	1.25	1.00
Manure spreader 130 BU	$3,042	1000	100	$527	$5.27	65	$1.98	$7.25	0.00	0.00
Trailer	$950	2700	200	$194	$0.97	60	$0.29	$1.26	0.00	0.00
Truck 1/2 ton	$8,700	3750	375	$1,526	$4.07	90	$6.50	$10.57	0.00	0.00
Tractor 25-hp	$8,776	6000	600	$1,464	$2.44	90	$3.93	$6.37	0.00	0.00
Tractor 35-hp	$11,791	6000	600	$1,962	$3.27	90	$4.82	$8.09	0.00	0.00
Tractor 40-hp	$13,475	6000	600	$2,238	$3.73	90	$4.56	$8.29	0.00	0.00
Tractor 50-hp	$16,136	6000	600	$2,676	$4.46	90	$5.59	$10.05	0.00	0.00

Source: Charles D. DeCourley and Kevin C. Moore, EC 959. Selected Fruit and Vegetable Planning Budgets. (University of Missouri–Columbia, Department of Agricultural Economics, 1987).

budgets and obtain such materials. We do not know of other sources of this kind of information. You will need to find sources for local genetic-stock seeds and plants. Many nurseries broker seeds and plants from outside their areas, even from overseas, and do not pay particular attention to the genetic origin of the plant products they sell. Take the time to find reliable sources, and avoid using seeds or plants from unknown locations. If nothing else, begin collecting your own local material and develop seed plots that can allow you to scale up when you can.

Our rule of thumb is that the seeds and plant products should come from the same physiographic region or natural area division (see Natural Area Division Maps for the USA, obtainable through most state or federal government agencies, or NaturServe). For insect-pollinated plant species, there should be even closer origins, no more than 100 miles. Physiographic region is fine for wind-pollinated species. If a nursery cannot tell you and guarantee a suitable genetic source of their plants and seeds, look elsewhere.

Contract growing as an option. Many nurseries can grow your plants and seeds with a year or so of advance notice, so they are ready when you need them. In fact, one way to reduce your purchasing costs is contract with a nursery to supply the materials. Because of the business certainty, a preorder allows them to reduce the unit costs compared to normal retail or even wholesale rates.

Contract growing is also a good way to ensure that you get local genetic stocks. Because many nurseries are only pass-through brokers and do little if any growing of their own plants and seeds, it is important to specify the genetic sources you require, and have a trusting relationship with the nursery.

On large projects, or projects where the added costs can be absorbed, a good nursery may actually collect local seeds in the vicinity of the restoration project to start nursery beds for producing the seed or plants required. This is ideal for projects where the local genetics really matter, such as a restoration to enhance or expand an existing nature preserve, national park, or state scientific natural area.

Partnering to produce seed. On some projects, companies such as AES and their nurseries, Taylor Creek Restoration Nursery (Great Lakes Region), Kaw River Restoration Nursery (Great Plains Region), and Spring Lake Restoration Nursery (North Woods Region), have partnered with the landowner by collecting and propagating local seed that has been planted into nurseries or fields in the client's restoration project site. Under joint venture agreements, AES has undertaken all responsibilities to plant, propagate, cultivate, weed, maintain, harvest, and clean the seed from such operations. Then, seeds needed for the restoration are available, and any excess is sold to other projects to the financial advantage of the joint venture. For very large restorations, creative relationships such as these can substantially reduce costs and ensure a supply of the highest quality local genetic seeds and plant stocks for your project.

On a cautionary note, we have seen restoration projects attempt to grow their own seeds with poor results. Even the easiest species to propagate often do poorly without the care and attention to detail that experienced nursery managers can provide. Many have found that growing many native species is challenging, but even more so is the harvesting and seed processing. On very small scales, the problems are manageable, but on larger scales, they can become overwhelming. For example, a normal agricultural combine cannot be used to harvest native seeds because most do not flow like soybeans or wheat for which combines were designed. Especially difficult are the fluffy, low-density seeds like asters, goldenrods, or grasses. A combine that has not been modified is useless. Steve learned the hard way, and partnered with experienced equipment operators who knew how to make the modifications necessary to allow the equipment to be used with native seeds. With native seeds, nearly constant adjustments also are needed as the humidity changes throughout a day of harvesting. For larger projects, it usually is most practical to partner with someone doing native seed production, especially for harvesting and cleaning.

You cannot legally use, donate, or share seed with others if it is not cleaned according to state and federal seed-labeling laws and requirements. It is beyond the scope of this book to explore the regulatory laws, but suffice it to say that it is illegal to remove from the premise of harvest any seed that does not meet the federal and state requirements for cleanliness. Generally, these laws were not written for application to native seeds. For example, if you harvested big bluestem grass (*Andropogon gerardii*) seed from plantings on your land, it would be illegal to share seed with a neighbor for their restoration if the seed included more than 1 percent inert matter (stems, insects) or more than 1 percent weed seed. Weed seed is defined as anything that is not the species on the bag label, including any native species that might be desirable in a restoration. The law requires that individual species be bagged separately. Therefore, seeds must be mixed after delivery at a restoration project site, although species can be mixed ahead as long as the label includes all the species in the mix.

Finding reputable suppliers. As in any industry, there are reliable suppliers and those with whom you would prefer not to work. Winnowing through a list and deciding is not an easy task. While most regional lists of native plant nurseries (available from natural resources agencies) are long, they include many who are unreliable. What is advertised by a nursery is often not delivered. In one project Steve ordered nearly 20,000 pounds of viable seed (called *pure live seed* [PLS]) and found that what was delivered only averaged 11 percent viable based on independent laboratory testing. When he attempted to return the seed, the vendor refused to accept it. That experience was why Steve's company (AES) started Taylor Creek Restoration Nursery.

Discovering reputable nurseries that will work with your specific project is primarily through word of mouth, based on the experience by others doing restoration work. Ask for references and check them out. Even then, it is best to meet with suppliers to discuss your needs, and conclude for yourself if their business philosophy is a good match to your restoration goals.

Hiring contractors. For several reasons, we recommend establishing the nursery source before negotiating with a restoration contractor for any services. First, if you supply the seeds and plants, this reduces the markup the contractors will place on those products if they purchase them. There are many good nurseries that also do contracting, and in those cases, you may be able to negotiate more favorable bids that include both the plant products and contracting services. Be sure to ask them to document the savings associated with providing both the plant materials and contracting services, and do comparison pricing. If a reliable source offers both the products and contracting, this also can save considerable time in looking for separate sources.

As your restoration plan is completed, you will begin to see clearly where you need to hire professional help. Determinants include needed equipment you lack and may not want to purchase, your budget, and needed expertise. Even if you want a turnkey restoration and can afford it, we strongly urge that you proceed through the planning steps to the point of implementation. After all, it is your land, and you will live with the results. You would no more want to turn over a restoration project to an outsider who does not know the land as you do than you would turn over the design of a house to someone who does not know your needs and desires. Moreover, you cannot make a good decision on what contractor to hire unless you have a very clear understanding of the job that needs to be done. There are two primary contracting strategies: design-build, when you hire one contractor to do or oversee all tasks in the restoration, or direct hiring, when you obtain bids from contractors for specific tasks during restoration.

A design-build contractor usually provides most or all specified restoration contracting services and materials, and subcontracts for what they cannot do themselves. Typically using a design-build contractor will save you 10 to 20 percent because the contractor controls the entire process. If you have a reliable contractor, this often is the best choice, especially if you have little equipment or experience. A reliable contractor usually assures quicker completion, and can result in fewer coordination issues (such as if you had multiple contractors involved).

Other things a design-build contractor will either handle or assist you with include the following:

- Help you review and finalize plans, principles, goals, and objectives leading to final design
- Get wholesale or price reduction for purchased materials and services
- Check reputations and assume responsi-

bility for the performance and cost/quality control of subcontractors
- Negotiate payout schedule
- Assume responsibility for the project (except acts of God)
- Provide guarantees and warranties
- Assign a project manager to oversee day-to-day work.

Long-term relationships benefit ecological restoration. Getting a good design-build contractor means that both parties are committed to solving problems that arise, and the contractor is committed to the overall project performance rather than simply doing a particular job. The project will benefit from a durable, long-term relationship, rather than a task-to-task one, involving many people coming and going.

It is easy to check a design-build contractor's proposed pricing. In many AES design-build projects, an independent third party has been hired to do price checking, or the contractor can be required to obtain bids for some services from some other contractors to demonstrate that their numbers are, in fact, competitive. Design-build contracts give nearly full responsibility for performance to the contractor. They are bound to perform and provide warranties and guarantees to this effect and bond or insure against "acts of God."

Of course any contractor is only as good as their communications. Make sure you are comfortable with who will work with you throughout the process. Unless it is with a pretty small contractor, chances are good that someone else actually will be on the job—thinking, affirming accountability, keeping you informed about what is happening, and openly discussing the challenges encountered. With any contractor, insist on a project manager or job foreman with whom you feel comfortable. Do not be victimized by someone who does not communicate well or does not give you the attention you deserve. Alternatively, be mindful of their time commitments and affirm to each other your expectations during the restoration process so you remain aligned.

The alternative to a design-build contractor is for you to act in that capacity and line up your own subcontractors, supplies, and equipment, as needed. On small tasks or projects, this approach may actually give you less flexibility in adjusting your implementation plan as needed, as any change may result in changes in pricing by the subcontractor. To bring the highest level of price certainty, however, you need to do the following:

- Prepare specifications and specific detailed scope of work for each subcontractor
- Bid out jobs, check references, and complete contracts
- Assume responsibility and liability for work and workers

When you hire subcontractors, you often must secure your own materials. To obtain accurate competitive bids, you will need a detailed scope of work for each task. Often, restoration plans lack adequate details, so we encourage you to work through the planning steps outlined in previous steps before you think about contracting for services. Especially if you do not provide a detailed scope of work, competitive bids can vary widely, depending on assumptions contractors make about what needs to be done. If bids are wildly dissimilar, you have no way of knowing why they are different. Do not leave this confusion unaddressed. This too often leads to poor results.

That said, there are some tasks that are easily and routinely defined for which you or your general contractor can get good competitive bids. A good example is earth moving. A grading plan with earth quantities is easily understood and can

be accurately bid. Bids often vary, however, according to how hungry a contractor is and how far they have to move equipment. Do not be surprised if you get a wide range of bids for even straightforward jobs.

For anything other than minor jobs, we recommend that you require performance bonds for services, warranties, and guarantees, and that these be legally binding between you and the contractor. Because this may affect bids, these requirements should be stated in the scope of work. Also obtain warranties from plant products suppliers. If they cannot obtain bonding or provide sureties, you should continue looking for someone who can. Because there are many inexperienced restoration companies and nurseries unfamiliar with good business contracting, we provide example templates for you to use with contractors or suppliers.

In appendix 3 we provide examples of standard contracts, which you may find useful. Each of the following documents can be customized for your particular project:

- Model design-build contract
- Model fee-for-services for a competitive bid process
- Performance bonding procedures
- Warranty language
- Seed and materials purchase agreement
- Consignment growing for seed and plant contract

Task 26. Complete Final Budget

You now have all the information to complete a final budget. Return to the master budget data form 4.2 (appendix 1) you began in step 4, task 17, and complete it with the additional information summarized and assimilated through step 7.

Task 27. Develop Long-term Maintenance Budget and Endowment

Some restored lands will not need much future maintenance, while others will require regular maintenance. Do not accept the opinion voiced by many that restored ecosystems will take care of themselves. Long-term funding depends to some extent on the goals and objectives you developed in step 4. This is a good opportunity to go back and revisit some of your goals, assumptions, and the guiding principles you developed.

Very few restored ecosystems can remain healthy without at least some maintenance. Unless your land is directly connected to large wilderness with intact ecosystems that have not been greatly altered by human activity, your project will always require some level of maintenance. If you do not make financing arrangements for maintenance, your heirs may feel they have higher priorities. If you gift your land to a land trust or agency, their good intentions for maintaining the long-term ecological health of the land may be overwhelmed by budget constraints.

To assist with this aspect of planning, we provide an endowment calculation tool (see table 7.2), which we will now walk you through its use.

You will need to create a ten-year budget for the monitoring, education, and on-the-ground maintenance activities you expect on your land after the first five years of restoration establishment. Because the endowment calculator is focused on year 6 and beyond, it includes costs only after the remedial restoration treatments have been successfully completed. Success means that any plantings are growing; ditch backfills are stable, and native vegetation is established.

For Stone Prairie Farm, Steve included his annual real estate tax in the budget to ensure that this was also considered in design of the endowment needed to annually generate the money to pay for all maintenance operations. Steve also assumed that his family would be doing all the annual monitoring to keep the family connected to the land. As a result, the costs for monitoring are simply some basic costs to feed and crunch data that are collected annually, rather than the additional costs if he were to contract out these monitoring services.

For each line item, estimates of the annual acreages and dollars needed are inserted into the table. The initial first-year costs (after year 5 of the "remedial restoration phase" is completed in each restoration zone) are escalated annually using a 5 percent cost inflation multiplier to make sure costs/prices are adjusted. A market-rate adjustment could also be conducted on some future schedule. However, by including this escalation in the budget, you can simply adjust the rate of escalation in the spreadsheet, perhaps on a five-year timeline.

This calculation estimates the endowment investment (e.g., annuity structure, or some annual dividend) that will generate the amount of money you need to fund the annual monitoring, maintenance, and any education programming. Your budgeting, of course, will require annual review as investment yields and expenses change over time. We suggest you review and revise the budget and endowment estimates at least annually. To revise the budget and endowment, insert the actual costs incurred during the restoration implementation process. Do not forget to sequentially number or uniquely date each revised budget so you never get confused about the history of the budgets you have developed.

You may still have remedial restoration in progress six years after starting. If so, your maintenance costs will become less as you complete them, and you need to adjust the inputs in

TABLE 7.2.

Perpetual maintenance cost and endowment calculation tool. Unit costs are used to calculate annual costs.

Project: Stone Prairie Farm

By- Steve Apfelbaum

	Qty	Unit	Cost per unit $	Year 1	Year 2	Year 3	Year 4	Year 5	Year 6	Year 7	Year 8	Year 9	Year 10
Burning	80	AC	10	800	840	882	926	972	1,021	1,072	1,126	1,182	1,241
Brushing	1	AC	2,000	2,000	2,100	2,205	2,110	2,216	2,327	2,443	2,565	2,693	2,828
Herbicide Weed Management	2	AC	100	200	210	221	232	243	255	268	281	295	310
Monitor Vegetation	80	AC	2	160	168	176	185	194	204	214	225	236	248
Monitor Hydrology	80	AC	2	160	168	176	169	177	186	195	205	215	226
Monitor Birds	80	AC	2	160	168	176	185	194	204	214	225	236	248
Macroinvertebrate Study	80	AC	1	80	84	88	93	97	102	107	113	118	124
Compile Monitoring Reports	1	YEAR	300	300	315	331	317	332	349	366	385	404	424
Brochure	1	EACH	1,200	1,200									
Open Houses and Neighbor Guides Hikes	1	YEAR	500	500	515	530	546	563	580	597	615	633	652
Trail Maintenance	1	YEAR	500	500	515	530	546	563	580	597	615	633	652
Educational Program	1	YEAR	300	300	315	331	317	332	349	366	385	404	424
Real Estate Taxes on Land	1	YEAR	2,000	2,000	2,100	2,205	2,110	2,216	2,327	2,443	2,565	2,693	2,828
Total				8,360	7,498	7,851	7,735	8,101	8,484	8,884	9,305	9,744	10,206
Total cost for 10 years	$86,169												
Averaged cost per year	$8,61												
Endowment required	$143,616	Assumes $8,616	Assumes $8,616.93	6.00%	APR								

the endowment calculator. This adjustment is needed on every project where phasing has resulted in the maintenance actually occurring in smaller acreages annually, at least until all remedial restoration phase activities are completed. With consolidation of maintenance over larger acreages, you should be able to find some cost efficiencies that may reduce the required endowment. You may also find that education needs differ from your expectations.

Summary

If you need plant materials or contractors for your restoration project, you should consider the important steps early in planning process. It is essential to interview and get comparative pricing if you are using costs and competitive bidding to determine who to hire. However, the best strategy, especially for large projects, often is a design-build firm that can help you finalize your plans, budget, organize, procure plant materials, and either do or oversee all subcontractors (e.g., earthmovers, regulatory experts, etc.). Design-build contractors commonly provide the highest quality outcomes at the lowest cost. In going this way, you can afford to invest more heavily in checking

credentials and reputation of potential contractors as you will be developing long-term relationships with them and relying on them for most of the success of the restoration. This also gives you more time to be overseeing and monitoring, reviewing results, and refining your restoration plan, discussed in the remaining three steps.

Planning is often the most difficult and time-consuming part of ecological restoration. If you start a restoration prematurely without considering these steps however, you can anticipate setbacks and extra expense. By thinking carefully through each step, you can avoid most of the potential problems. You now have a completed budget and a good idea of the cash flow requirements for implementation as well as some long-term projections for maintenance.

You are now ready to implement your restoration. Follow the schedule and initiate tasks as planned, but be prepared for the unexpected. As many have observed, nature can be fickle, and no plan is perfect. Regardless of whether things run like clockwork, or you have to revise your plan frequently, record the details of your work on the land and the responses. We explain in the next chapter how good records help you refine your restoration plans and implement the program.

Maintain Good Records

There is no more difficult art to acquire than the art of observation, and for some men it is quite as difficult to record an observation in brief and plain language.

William Osler

Records of monitoring are essential, but records should go much further, with details of the restoration process. Restoration involves manipulation of many variables, with the aim of urging ecological systems toward restoration goals and objectives. Without records, even someone with an excellent memory will soon become vague in what practices were employed, and when.

Record keeping involves (1) recording where and when work was completed spatially on the plan maps, (2) recording the time and costs to do the work, and (3) successes and failures as indicated by monitoring. See fig. 8.1 as an example, using a prescribed burn task. Similar records should be compiled for every task.

We also advocate record keeping of communications with neighbors and others who might be involved in the restoration process. Who expressed concerns, who visited and used the land for educational use, for hunting, for training, all become important in telling your restoration story within your family and to neighbors and others in the future. For example, if smoke from prescribed burning concerns a neighbor, choosing a day when the wind carries the smoke away

from the concerned neighbor's home might be possible, but only if your records reminded you of that concern.

Restorations can get attention in your neighborhood and community. In the restoration planning worksheet, we encourage you to keep a running record of your external communications about your project. Steve initially failed to record the names and contact information of the hunters who knocked at his Stone Prairie Farm door, asking if they could hunt on his restored prairie. In retrospect, it would have been useful to contact them to share some educational information about the restoration, and why the wildlife responded so positively. Steve began summarizing conversations with neighbors. Over time, these conversations revealed who was entertaining ideas about restoration of parts of their farms, and this gave rise to cooperative projects.

Stories about the history of the land, hunting and trapping, fish "grandfather" used to catch in the neighboring creek, and when the railroad came through begin to tell a lot about the landscape before it was so developed. At a neighborhood Christmas party, Steve learned that his farm

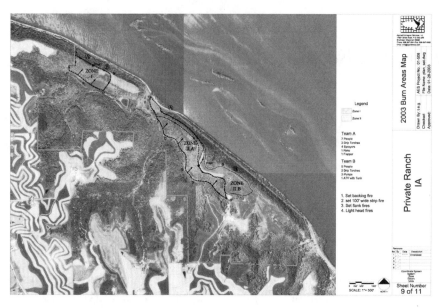

FIGURE 8.1. Example of record of a prescribed burn at a private ranch, IA

used to support stands of sandbar willow (*Salix interior*) along the spring brook, and that it was where a neighbor's grandfather in the late 1800s could nearly always shoot a few prairie chickens on a cold winter morning. Because prairie chickens had been extirpated by 1948, this was an exciting tidbit of history.

Photographic records of pretreatment, restoration work, and responses are one of the best ways to convince reluctant stakeholders that a project is worthwhile. Showing the transition, such as converting a manicured farm field to native prairie, with the predictable weeds the first few years, has persuaded many discouraged landowners to persist in their restoration work. Images of test or demonstration plots showing responses to thinning the canopy in dry forests with regrowth of native grasses and wildflowers makes a compelling case for restoration. The visceral reaction of most people to any tree cutting is often

negative. Photographs from projects can help convince reluctant stakeholders. Complete records of activities associated with restoration are useful and necessary for the following reasons:

- Precision and accuracy in conveying understanding and knowledge
- Planning operational details enriched with prior experience
- Transferring responsibilities between parties contributing to restoration
- Maintaining an engaged and informed restoration team
- Demonstrating results from past investments to leverage new funding and involvement by others
- Documenting the story

There are many kinds of information that might become important in evaluating the response to treatments on the land, or to help you revise your plan based on accumulating experience:

Labor: time and effort invested. It is often useful to segregate volunteer hours from contracted hours, as these can sometimes be used as a match in grant proposals.

Finances: services hired or volunteer equivalency invested in the project, and material purchases, equipment costs, fuel, and other expendable costs. Equipment amortization may also be useful, if you have your own.

Planting materials: be sure to keep good records of seed sources, quantities and qualities, or ages and sizes. In addition to site preparation, the choice of planting materials often determines success or failure.

Dates when work was done: planting too late or too early can affect results, but more important is recording the timing in degree-days or measures that relate to soil moisture or seasonal temperatures. Weather surrounding restoration activities is usually important.

Monitoring data: formal monitoring data are usually tied to specific treatments or management units and a sampling protocol. Less formal monitoring involves observing and recording notes and photographs. The data are commonly key to understanding conditions and responses to restoration treatments.

Observations and thoughts: regular entries into a journal about what you see, how you think things are working, and even how you feel about those things can lead to broader insights. Observations of others who are involved with a restoration also can be helpful, so all should be encouraged to record observations, dates, times, and events on the land. This is a great way to get children involved in observing and writing. Consider a family journal in which all are encouraged to participate.

On-ground photographs: these are so easy with digital cameras that the real challenge becomes dating and georeferencing the images so that future photographs can be matched. Without these data, photographs are of limited use. Many projects use permanent photographic stations to document conditions repeatedly, especially to document the responses to restoration treatments. The downside of digital images is that massive numbers of images can be inexpensively collected and archived, increasing the demands for good records if they are to be useful.

Aerial photographs: we are converting globally from hard-copy aerial photographs to digital images. Obtaining, organizing, and archiving paper or digital aerial images is one of the easier types of record keeping because the source provides dates and other salient aspects of the images. Now, there is also very high resolution, multispectral digital images available (see Applied Ecological Services website [www.appliedeco.com]) with ground resolution of one inch or less. This means you can use this high-resolution imagery for measurement of vegetation cover, among other variables.

Phenological records: Aldo Leopold's family continued his practice of recording first dates of observations, such as blooming of wild flowers and arrival of migratory birds. Because of the long-term record, their notes have become an important database that documents climate change. While these records may tell you relatively little about responses to restoration treatments, they help you focus on new species that show up, or changes in those already present.

The importance of record keeping often slips by people who think they have great memories, or those who simply believe they have only enough time to get restoration tasks completed. In reality, record keeping is as essential as any restoration task when viewed in hindsight, four or five years into a project.

When record keeping is left up for grabs, nobody does it, and there is a loss of learning from the successes or failures in the project. Both the simple details of when and how, as well as finer details should be noted. Results of monitoring can become useful only when related to this kind of information, and adaptive management will be possible only with both.

Task 28. Decide What Will Be Recorded

The easiest way to organize records is spatially and temporally. Recording your data spatially means tying the records, including measurements, observations, or thoughts, to locations in the landscape, typically management units. Noting the date of data collection or when observations were made is an obvious requirement. Digital records can be stored and retrieved by date or location.

Both the date and geographic locations can be easily recorded using the graphic restoration plan that you have created. Each time you want to make a record, you can generate a new plan sheet on which you note dates and kinds of restoration activities, observations, new plant or animal records, or other information. Parallel files of monitoring data can be cross-referenced from the plan sheet. You should also keep a photographic log with the images numbered and briefly described (see steps 2 and 6). For each image, enter the number with a small arrow showing the direction or orientation of the images. Some digital cameras now have an internal GPS and will auto-number and allow you to download the images and the location where you took each photograph, time of day, and date. These cameras will also provide the alignment arrow automatically, showing the direction of the image.

Digital or paper? You usually will carry a copy of your basemap into the field, perhaps along with a hand-held GPS unit and a digital camera. In addition, you should also carry a small journal or notebook. In the field, record brief observations and thoughts on the basemap, supplemented with photographs and more lengthy inclusions in your journal. All photographs and journal entries are GPS located when appropriate, and it is even convenient to note GPS location of entries on the basemap to facilitate relocating that point in the future.

If you have digital-mapping capability, the information from the field can be saved to your computer. We like to maintain a computer record of our journal entries, as well. Each set of observations or data becomes an additional layer of information on the digital map and can be accessed by date or location, including the management unit.

Ready access to paper maps is important. Develop a strategy that makes it as easy as possible to keep records. The more often you make observations, and the more information you record, the greater your insights and understanding about the responses to your restoration treatments. Being able to race out the door with a copy of the basemap and a pencil facilitates a good record of what is happening on the land. Keep a field journal and extra copies of your basemaps handy. Come a rainy day, digitize the stack of marked-up basemaps. We suggest creating twelve hard copies of the basemaps each year and using one a month. This allows you to remain focused on learning about the changes in the land rather than processing data. Don't neglect your journal. Steve's collection of thirty years of journal records was the basis for his book, *Nature's Second Chance*.

Task 29. Develop a Record-keeping Strategy

Journaling and photography: It will be obvious how to keep most information once you decide what you want to keep and how it will be organized. You should not overlook, however, journaling and photography.

Journaling can capture the essence of the restoration through your observations and reflections, even if only after some important events—planting the prairie, the return of bobolinks, or seeing a bobcat. Whether you write or not, it is important to take the time to reflect on your observations. For most of us, writing aids this process. It helps put the observations of the day or week within a context of place, history, and changes occurring on the land.

While baseline and posttreatment data may

speak to scientific minds, nothing can convey some changes that occur with restoration better then photographs. We emphasize the importance of being able to return to the same locations seasonally, annually, and at special times (e.g., after a snow storm, a prescribed burn, ice storm, drought, etc.) to document how the ecosystem has responded to the restoration treatments or to natural events. Establish permanent photograph stations for your project during the baseline sampling process, and capture prerestoration conditions to compare with posttreatment photographs taken at regular intervals.

If you do not use GPS, metal fence posts can be used to establish these photograph stations. With either method, there is a need to orient the camera the same way each time. A compass can be used along with a note or two in your photographic record about height and direction and light conditions. A metal fence post can be oriented with the rib side of the post facing the scene, and the top of the post can be where you place your camera.

Results of restoration are most simply captured with before and after photographs. These images often provide personal insights into how a landscape or stream changed with restoration. Because changes are often slow and gradual, they often do not make as much impression, especially when we are see them day after day. Recall the shock when viewing photographs of yourself or family taken over the years and the contrasts the photos reveal.

A sequence of images showing the changes over time is a great way to share the restoration experience with others. When the images are provided in parallel with monitoring data, particularly vegetation composition and structure data, the richness of the restoration is revealed. Add tidbits of your personal essays, and you will have created a record that is entertaining, enlightening, and inspiring.

Summary

Collecting and keeping good, meaningful records that include technical data, written reflections, and good photographs captures your legacy on the land. Furthermore, what was done, what worked, and how the land responded guides future efforts and your opportunity to learn from your experience. Relying only on your memory will not be of great help for you or others who try to understand the experience or avoid most mistakes in the future. Your project and story of restoration will be an important incentive to encourage others to restore their land. Never discount the value of your records.

Step 9.

Review the Project

Any activity becomes creative when the doer cares about doing it right, or better.

John Updike

We increase success in just about anything we do if we schedule regular reviews to adjust for any changed conditions and critically examine progress. When objective third parties can be included in the process, even more can be gained. Reviews should be open, candid, and honest. Good monitoring data and good records are a primary basis for good reviews.

The process of using performance information to review and refine the restoration program is called *adaptive management*. Design and schedule reviews so that adaptive refinements can become part of the evolving plan, not just a side process that does not inform the overall program. Refinements may also be required as a result of new conditions such as changes in financing, delays resulting from labor shortage, or weather.

Some tasks must be timed to coincide with specific biological and meteorological events. For example, if you are planning on using prescribed burning, make sure you have included a task to obtain any necessary approvals or permits ahead of time. With that scheduling task, make sure you realize the best time or times of year when burning can be done. At Stone Prairie Farm, burning can be conducted in early spring before vegetation greens up, during the heat of the midsummer if there is a dead thatch layer underlying the green vegetation, or in the fall after all vegetation becomes dormant. Steve scheduled spring and midsummer burns to better control cool season grass growth and invasive woody vegetation, respectively. Often, however, conditions have not been suitable, and alternative autumn burns have been done. The point is, you may have to reschedule some tasks, perhaps at less optimum times.

Figure 9.1 emphasizes there is a season or time for many tasks. Align the tasks that must be seasonally or phenologically specific with the scheduling of other tasks. This will include not only prescribed burning, but weed management (as some techniques only work at specific times), seed collecting, and planting. Monitoring schedules also are time sensitive. Monitoring to compare changes over time, from year to year must be completed at about the same time each year.

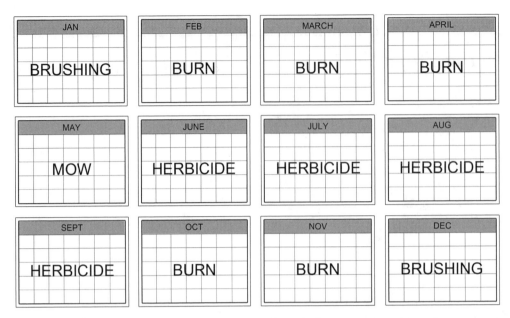

FIGURE 9.1. Schedules for restoration projects must coincide with the appropriate times of year

Task 30. Schedule Semiannual and Annual Reviews

Schedule a semiannual and an annual review with the immediate project team (e.g., your family, your land trust members, or others) and, if possible, one or more people outside the immediate team. Include reviews in the schedule of tasks for each year. To get the most out of the review, restate goals and objectives and summarize the data available to assess progress toward them. Distribute these to the team at least two weeks prior to the meeting. The most useful data are those that provide information about the changes attributable to treatments, but don't overlook evidence such as aerial photographs and data on invasive or rare species abundance. If you have the capability, there are many types of graphs you can easily create to give reviewers a quick understanding of the trajectory. Keep in mind that out-

side reviewers will be much less informed about methods and treatments, so a thorough review should include at least a summary of what was done, how, and when.

We also recommend a field visit to inform the review team members, including persons who may have even been actively involved in the day-to-day work in implementing the restoration plans. There is no better way to learn and grow together under a common understanding than by sharing a field visit to discuss each member's perspective on what has occurred in representative field settings in a restoration project.

Task 31. Refine the Restoration Program as Suggested by the Review

Often described as *learning by doing*, adaptive management is essential when working with systems where outcomes are uncertain. Adaptive management is an iterative process, beginning with hypotheses or models of how you expect the system to behave (step 3). With restoration, your hypotheses are based on your understanding of how the ecosystem functioned historically and how stressors are affecting them now, as well as what you expect to result from each specific treatment. Monitoring (observations and data) are examined to determine if hypotheses can be accepted and the model confirmed. Usually there are variations, and the analysis leads to refinements in your understanding of how the system works. This is the learning part of adaptive management. With a better understanding, you develop new or revised hypotheses and models, initiate alternative treatments, continue monitoring and analyzing results, and so on. With complex ecosystems, you will never gain complete understanding. Moreover, ecosystems are continually changing. Thus, with adaptive management you may reach a point where you are reasonably comfortable with your knowledge and ability to predict outcomes, but there will always be a degree of uncertainty.

Plan to spend some time toward the end of the review process discussing what your observations and data mean about how ecosystems have responded to the various treatments. If possible, reach consensus on what you have learned and how that learning can best be applied in refining the restoration plan. This is usually the most exciting and certainly the most rewarding part of the review process, and those you ask to participate in it should be afforded the opportunity to share in that excitement. Capture the new knowledge with recordings or notes, and promptly summarize it to distribute to participants within two weeks, while it is still fresh in your mind. Be sure to schedule the review when you will have time for this follow through. The ideal time is usually late winter, when all data from the previous season has been compiled, and well before final plans are made for the next season.

Share the Restoration Process

We can teach from our experience, but we cannot teach experience.

Sasha Azevedo

Sharing what you learn with others can become a most important outcome of your restoration project. Your experience in developing and implementing your restoration provides you with a wealth of knowledge that should be widely shared. Both your successes and failures will be useful to those interested in restoring their land. Also, your commitment, enthusiasm, and good example are likely to inspire others to follow your lead. As Margaret Mead once observed, "Never doubt that a small group of thoughtful, committed citizens can change the world; indeed, it's the only thing that ever has."

Task 32. Develop a Plan for Sharing Your Experiences

We recommend that you identify three venues annually, in which you can share your experience, from the simplest understandings of the process to profound insights, if they occurred. We have found that scanned images of the overlays you developed and used to design your restoration program coupled with observations and photographs make a great story. The opportunities for sharing are nearly endless, from garden clubs, bird clubs, and conservation organizations to school classes. Neighbors will be curious about what you are doing and why. Results usually excite people, especially if you can share your enthusiasm. The thrill of seeing a bobolink return to a restored meadow on a spring morning can stir the heart of old and young alike. You need only strive to help your audience understand the process of restoring nature, an inevitably exciting subject for a wide range of people, including church groups.

Of course there also are opportunities for technical presentations to professional and scientific groups or conferences. Do not doubt that your experience represents important knowledge that professionals will want to hear about. Depending on the nature of the audience, the subject matter can be customized to tell the whole story or any part of the restoration program. We have given dozens of lectures to professional audiences and published several professionally reviewed papers based on nothing more than monitoring data captured from various restoration projects.

One of the best ways to share the experience is through fieldtrips. These can be entirely focused on explaining the process, or on jobs where a few extra hands are useful. Collecting seeds, sampling streams, or doing bird or butterfly censuses are examples where classes, organizations, or volunteers can be engaged in the process and learn by doing while helping you.

You may want to develop photo albums in which you put selected photos with appropriate captions to show the process and results. These will be of interest to family, friends, and neighbors, but also can be used in presentations when converted to slides. Excerpts from your journal can enrich the album and your presentations. If you have younger children, get them interested in contributing or perhaps doing their own album, and what a great "show and tell" story they will have to share with their classmates.

Working closely with nature is rewarding and exciting. Indeed, it may be the most important reason for doing restoration. Share the reward.

Appendix 1.

Data Forms

Each chapter references data forms to facilitate your restoration work. All forms are compiled in this appendix. The data forms have been numbered to help you locate where they are referenced in the text. For data form X.Y, X represents the chapter where the data form is referenced, and Y represents the number within that chapter. For example, data form 4.1, "Goals and Objectives," is the first data form referenced in chapter 4, and data form 1.5, "GPS Record," is the fifth data form referenced in chapter 1.

As you work through the sequential steps described in the restoration process, begin filling out the appropriate data forms. We recommend that you scan and print, or photocopy the data forms, creating multiple copies to have some extras on hand, some for scratch sheets and others to share if needed.

All data forms have been designed to be easily customized to meet the needs of your specific project (s). Some data forms can be completed as you read through questions in the respective chapters where the data forms are first referenced. Should you have questions about how to complete the process and entries in the data forms, please refer to the respective chapter where instructions are provided.

Data form 1.1: Restoration and Management Planning
Data form 1.2: Field Notes
Data form 1.3: Catalog of Features
Data form 1.4: Photograph Record Data Form
Data form 1.5: GPS Record
Data form 1.6: Sample Collection
Data form 1.7: Species Listing
Data form 1.8: Ecological Change Reasons
Data form 1.9: Soil Data Form
Data form 1.10: Stream Channel Data Form and Drainageways Mapping

Data form 1.11: Invasive Plant Species
Data form 2.1: Historic Conditions
Data form 2.2: Oral History Interview
Data form 3.1: Gradient Analysis
Data form 4.1: Goals and Objectives
Data form 4.2: Master Budget
Data form 6.1: Herbaceous Vegetation Cover
Data form 6.2: Woody Vegetation Cover
Data form 6.3: Bird Point Count Plotting
Data form 7.1: Restoration Scheduling

Data Form 1-1, Restoration and Management Planning

RESTORATION AND MANAGEMENT PLANNING

I) Management Unit: _____ and Unit Name: _____

II) Existing Vegetation Type (Reference or attach description):_____

III) Historic Vegetation Type (Reference or attach description): _____

IV) Noxious weeds present that may need focused management attention:

1. _____   4. _____
2. _____   5. _____
3. _____   6. _____

V) Significant change in **physical** (hydrology, drainage, excavation, dredging, sedimentation or erosion, etc.), **chemical** (contaminants, nutrients, erosion/sedimentation, agricultural/development, surface water loading, etc.), and **biological** (other problem species, seed bank depleted, shade suppression of ground cover, dominance by a few species, etc.) components of the unit:

1. _____   6. _____
2. _____   7. _____
3. _____   8. _____
4. _____   9. _____
5. _____   10. _____

VI) Restoration/Management Goals: {Note Priority or sequence of events}

{ } 1. _____   { } 6. _____
{ } 2. _____   { } 7. _____
{ } 3. _____   { } 8. _____
{ } 4. _____   { } 9. _____
{ } 5. _____   { } 10. _____

VII) "Monitoring Attainment of Goals" tied to goals above

	Parameter to Measure	Method to use	Timing/Frequency
1			
2			
3			
4			
5			
6			

VIII) Restoration/Management tasks* tied to goal above:
(*Property boundary surveying, landowner contacts, educational programs, brushing, noxious weed treatment, prescribed burning, install fences, install firebreaks, meetings, monitoring, photography, tours, volunteer efforts, urban wildlife management task, seed collection, planting, propagation, and reintroduction of species, press conferences and PR, research, review and refinement of plan, etc.)

1. A.
   B.
   C.

2. A.
   B.
   C.

3. A.
   B.
   C.

Data Form 1-1 continued

I) Phasing/Scheduling of Tasks:

	Remedial								Maintenance																Who does task	Days of effort required
	Q1	Q2	Q3	Q4	Q1	Q2	Q3	Q4	Q1	Q2	Q3	Q4	Q1	Q2	Q3	Q4	Q1	Q2	Q3	Q4						
	Year 1				Year 2				Year 3				Year 4				Year 5									
1A																										
B																										
C																										
2A																										
B																										
C																										
3A																										
B																										
C																										

II) Concerns that cannot be addressed by on-site restoration/management:

1. _____   4. _____
2. _____   5. _____
3. _____   6. _____

III) Site and off-site problems that need to be addressed or considered by others (i.e. land acquisition, property boundaries, wetland creation, permits, railroad, community outreach, volunteerism, etc.):

1. _____   6. _____
2. _____   7. _____
3. _____   8. _____
4. _____   9. _____
5. _____   10. _____

IV) Contacted persons, agencies: Names, address, and telephone numbers.

1. _____
2. _____
3. _____
4. _____
5. _____

V) Running record of notes of correspondence on the above: (date, person, point of communication)

1. _____
2. _____
3. _____
4. _____
5. _____

VI) Process to allow implementation of program: (political, education, cooperative and coordination issues, landscape ecology issues).

1. _____
2. _____
3. _____
4. _____
5. _____

Data Form 1-1 continued

IX. Phasing/Scheduling of Tasks:

	Remedial								Maintenance																Who does task	Days of effort required
	Q1	Q2	Q3	Q4	Q1	Q2	Q3	Q4	Q1	Q2	Q3	Q4	Q1	Q2	Q3	Q4	Q1	Q2	Q3	Q4						
	Year 1				Year 2				Year 3				Year 4				Year 5									
1A																										
B																										
C																										
2A																										
B																										
C																										
3A																										
B																										
C																										
4A																										
B																										
C																										
5A																										
B																										
C																										
6A																										
B																										
C																										
7A																										
B																										
C																										
8A																										
B																										
C																										
9A																										
B																										
C																										
10A																										
B																										
C																										

From *Restoring Ecological Health to Your Land Workbook* © 2011 Steven I. Apfelbaum and Alan Haney

Data Form 1-2, Field Notes

Observer: _____ Record Period:_____ Map Unit Name: _____

Record Identity *No/date*	Observation	Y/N Transferred to Hand Sketch
1- *date*	Observation:	Y / N
2- *date*	Observation:	Y / N
3- *date*	Observation:	Y / N
4- *date*	Observation:	Y / N
5- *date*	Observation:	Y / N
6- *date*	Observation:	Y / N
7- *date*	Observation:	Y / N
8- *date*	Observation:	Y / N

From *Restoring Ecological Health to Your Land Workbook* © 2011 Steven I. Apfelbaum and Alan Haney

Data Form 1-3, Catalog of Features

1. Discontinuities:
   - Books
   - Difference

2. Features:
   - A. Long Term
     - Rock
     - Bare Soil
     - Streams
   - B. Temporary
     - Bare Soil
     - Streams
   - C. Historic

3. Changes on the Land
   - Water ways
   - Erosion
   - Invasive young trees
   - Tillage, plowing, bull dozing

4. Remnant, Historic Conditions
   - Archeological
   - Abandoned, bulldozing, fences, etc.
   - Fire Scars, disease, insect infiltrations

5. Habitats
   - Terrestrial/Aquatic

6. Grass and Vegetation

7. Boundaries
   - Property
   - Use Borders

8. Land Management Features
   - Logging/Reforestation
   - Other
   - Plow
   - Orchards
   - Bare untilled soil
   - Wells, Water Management Features

9. Livestock and Wildlife Feature

10. Infrastructure
    - Transportation
    - Foods, Lanes, Bridges, Blasted Rock
    - Pipes, Siphons, Drainage tiles
    - Drainage/Irrigators, surface water diversions/controls
    - Electrical – wind, power line, electrical poles, generator
    - Quarries/Rivers/Borrow sites

From *Restoring Ecological Health to Your Land Workbook* © 2011 Steven I. Apfelbaum and Alan Haney

Data Form 1-4, Photograph Record Data Form

Photographer: _____   Location: _____

Photo Index. jpg#	Date	Description	Instructions	Repeat Status
				Y / N
				Y / N
				Y / N
				Y / N
				Y / N
				Y / N
				Y / N
				Y / N
				Y / N
				Y / N
				Y / N
				Y / N
				Y / N
				Y / N
				Y / N
				Y / N
				Y / N
				Y / N
				Y / N
				Y / N
				Y / N
				Y / N
				Y / N
				Y / N
				Y / N
				Y / N
				Y / N
				Y / N
				Y / N
				Y / N
				Y / N
				Y / N
				Y / N
				Y / N
				Y / N
				Y / N
				Y / N
				Y / N
				Y / N
				Y / N
				Y / N
				Y / N

From *Restoring Ecological Health to Your Land Workbook* © 2011 Steven I. Apfelbaum and Alan Haney

Data Form 1-5, GPS Record

GPS Operator: _____ Location: _____

GPS Point Number	Date	Description	Instructions	Repeat Status
				Y / N
				Y / N
				Y / N
				Y / N
				Y / N
				Y / N
				Y / N
				Y / N
				Y / N
				Y / N
				Y / N
				Y / N
				Y / N
				Y / N
				Y / N
				Y / N
				Y / N
				Y / N
				Y / N
				Y / N
				Y / N
				Y / N
				Y / N
				Y / N
				Y / N
				Y / N
				Y / N
				Y / N
				Y / N
				Y / N
				Y / N
				Y / N
				Y / N
				Y / N
				Y / N
				Y / N
				Y / N
				Y / N
				Y / N
				Y / N

Data Form 1-6, Sample Collection

Observer: _____   Location: _____

Sample Bag Record Number	Date	Description

From *Restoring Ecological Health to Your Land Workbook* © 2011 Steven I. Apfelbaum and Alan Haney

Data Form 1-7, Species Listing

Observer/Recorder : _____   Location: _____

Date	Species Observed	Date	Species Observed

From *Restoring Ecological Health to Your Land Workbook* © 2011 Steven I. Apfelbaum and Alan Haney

Data Form 1-8, Ecological Change Reasons

Question Number	Question	Location on Figure 3.2	Answer(s)

From *Restoring Ecological Health to Your Land Workbook* © 2011 Steven I. Apfelbaum and Alan Haney

Data Form 1-9, Soil Data Form

Location/GPS point:

Sample number:

Sampler:

Date:

1. Duff / litter layer

    Depth:

2. "A" topsoil horizon

    Depth:

    Texture:

    Color:

    Notes:

2. "B" subsoil horizon

    Depth:

    Texture:

    Color:

    Notes:

2. "C" parent material horizon

    Depth:

    Texture;

    Color:

    Notes:

Draw soil layer depths and characteristic features in rectangle below. Features identified may include:

-plow depth   -mottles

-saturation depth   -debris

(Depth in inches)

0
1
2
3
4
5
6
7
8
9
10
11
12
13
14
15
16
17
18
19
20
21
22
23
24

From *Restoring Ecological Health to Your Land Workbook* © 2011 Steven I. Apfelbaum and Alan Haney

Data Form 1-10, Stream Channel Data Form and Drainageways Mapping

Stream Name: _____

Sample #: _____

GPS Location: _____

Data Collector: _____

DATE: _____

Input data	Field Measured	Primary Calculations	Secondary Calculations
Watershed metrics (from USGS or Equi map)			
Tributary area (square miles)			
Channel slope (ftv/ft hor)			
Valley Slope (Ft v/ ft hor)			
Verticle drop over Valley slope (ft/ft)			
Valley Length (ft)			
Stream Length (ft)			
Channel metrics (Field measured)			
Bankfull width (ft)			
Bankfull mean depth (ft)			
Bankfull cross sectional area (ft 2)			
Bankfull Discharge (CFS)			
Mean stream velocity (ft/sec)			
Wetted perimeter (ft)			
Radius of Curvature (measured from aerial -Ft)			
Width/Depth ratio (Width Bkf/Depth (Bkf mean)			
Entrenchment ratio (Flood Prone width/Bankfull width)			
Sinuosity (Stream length/Valley Length) Ft/Ft)			
Water Surface Slope (Verticle distance (ft)/Linear Distance (ft)			
Meander Width ratio (Meander Belt width/Width at bankfull)			
Flood Prone area Width (Wfp=2 (bkf depth) projected horizontally			
Stream Pattern metrics (Aerial and ground measurement)			
Pool riffle distance (ft)			
Meander belt width (ft)			
Meander distance (ft)			
Substrates and bank Metrics (field measurement)			
Diameter of 50% substrate size (cm)			
Diameter of 84% substrate size (cm)			

From *Restoring Ecological Health to Your Land Workbook* © 2011 Steven I. Apfelbaum and Alan Haney

Data Form 1-10 continued

Input data	Field Measured	Primary Calculations	Secondary Calculations
CALCULATIONS MATRIX			
Radius of Curve calculations			
Meander length vs channel width	L=10.9W^ 1.01		
Meander Length vs mean readius of curvature	L=4.7 Rmean^ 0.98		
Bankfull mean width vs radius of curvature	D=.85Rc^0.66		
Radius of curvature vs bankfull cross sectional area	Rc=wmean/.85 *0.66		
Bankfull crosssection area vs radius of curvature	A=.067Rc^1.53		
Radius of curvature vs bankfull cross sectional area	Rc=A/.067*1.53		
Bankfull width vs radius of curvature	W=.71 Rc^0.89		
Radius of curvature vs bankfull width	Rc=w/.71 *0.89		
Shear Stress and Velocity Relationships			
Shear stress (Lbs/ft 2)=D x R x S			
D-density of water assumed=1			
R=hydraulic radius (wetted perimeter of channel)			
S=channel slope (ft/ft)			
Hydraulic Radius (R=A/WP)			
R=hydraulic radius of channel			
A=Cross sectional area (ft^2)			
W=Channel Width (ft)			
P=Wetted perimeter (ft)			
Roughness Coef at Bankfull stage			
N=1.486/Qbkf (Area*Hydraulic radius^2/3) (Slope^1/2)			
Mean Velocity			
V=1.49®^2/3 *(s)^1/2/n			
Bankfull stage hydraulic geometry calculations			
(using NC regional MTN curves)			
Bankfull discharge (cfs)	Qbkf=115.7 Aw^0.073		
Bankfull Area (ft^2)	Abkf=22.1 Aw^0.67		
Bankfull Width (ft)	Wbkf=19.9Aw^0.36		
Bankfull Depth (ft)	Dbkf=1.1Aw^0.31		

From *Restoring Ecological Health to Your Land Workbook* © 2011 Steven I. Apfelbaum and Alan Haney

Data Form 1-11, Invasive Plant Species

Observer/Recorder : _____   Location: _____

Invasive Record Number	Date	Species	# individuals/stems or estimated population size	Population Growth and spread (Stable / Enlarging/ Declining)		
				S	E	D
				S	E	D
				S	E	D
				S	E	D
				S	E	D
				S	E	D
				S	E	D
				S	E	D
				S	E	D
				S	E	D
				S	E	D
				S	E	D
				S	E	D
				S	E	D
				S	E	D
				S	E	D
				S	E	D
				S	E	D
				S	E	D
				S	E	D
				S	E	D
				S	E	D
				S	E	D
				S	E	D
				S	E	D
				S	E	D
				S	E	D
				S	E	D
				S	E	D
				S	E	D
				S	E	D
				S	E	D
				S	E	D
				S	E	D
				S	E	D
				S	E	D
				S	E	D
				S	E	D
				S	E	D
				S	E	D
				S	E	D

From *Restoring Ecological Health to Your Land Workbook* © 2011 Steven I. Apfelbaum and Alan Haney

Data Form 2-1, Historic Conditions

Mapable Answers	Date	Question	Answer(s)

Data Form 2-2, Oral History Interview

Interviewer : _____   Person Interviewed: _____   Date: _____

Questions Number	Answers and map reference location
1. _____ Initials	
2. _____ Initials	
3. _____ Initials	
4. _____ Initials	
5. _____ Initials	
6. _____ Initials	
7. _____ Initials	
8. _____ Initials	
9. _____ Initials	
10. _____ Initials	
11. _____ Initials	
12. _____ Initials	
13. _____ Initials	
14. _____ Initials	
15. _____ Initials	
16. _____ Initials	
17. _____ Initials	
18. _____ Initials	
19. _____ Initials	
20. _____ Initials	
21. _____ Initials	
22. _____ Initials	
23. _____ Initials	
24. _____ Initials	

From *Restoring Ecological Health to Your Land Workbook* © 2011 Steven I. Apfelbaum and Alan Haney

Data Form 3-1, Gradient Analysis

Type of gradient and corresponding ecotone	Description

From *Restoring Ecological Health to Your Land Workbook* © 2011 Steven I. Apfelbaum and Alan Haney

Data Form 4-1, Goals and Objectives

Stressors A. What i. Where	Goals	Objectives	Primary Restoration Tasks

From *Restoring Ecological Health to Your Land Workbook* © 2011 Steven I. Apfelbaum and Alan Haney

Data Form 4-2, Master Budget

Restoration Task	Unit	Quantity	Unit Cost	Cost ($)	Assumptions

From *Restoring Ecological Health to Your Land Workbook* © 2011 Steven I. Apfelbaum and Alan Haney

Data Form 6-1, Herbaceous Vegetation Cover

Date: _____

Plot Size: _____

Unit and Cover Type: _____

Data Collectors: _____

Page ____ of ____

Transect Info.

Transect #	1	2	3	4	5	6	7	8	9	10
Transect length (m)										
Orientation of transect										

Transect #																				Total	Average			Relative		Relative	Importance
Random Distance from transect (m)	R	L	R	L	R	L	R	L	R	L	R	L	R	L	R	L	R							Cover	Frequency	Cover	Frequency	Value
Quadrat #	1	2	3	4	5	6	7	8	9	10									N			Frequency	Average Cover				
Species	Cover	Cover	Cover	Cover	Cover	Cover	Cover	Cover	Cover	Cover																	

Litter										
Rock										
Bryophyte/Lichen										
Bare Ground										

Percent Cover Classes: 1 = 0-1%   2 = 2-5%   3 = 6-25%   4 = 26-50%   5 = 51-75%   6 = 76-95%   7 = 96-100%

From *Restoring Ecological Health to Your Land Workbook* © 2011 Steven I. Apfelbaum and Alan Haney

Data Form 6-2, Woody Vegetation Cover

TREE & SHRUB CANOPY INTERCEPT
Trees & shrub > 1m height

Project/Location:

Date:_____   **Transect:**_____   **Samplers:**_____

Species	Line intercept on 100 meter line		
		Total	Means

From *Restoring Ecological Health to Your Land Workbook* © 2011 Steven I. Apfelbaum and Alan Haney

Data Form 6-3, Bird Point Count Plotting

Date: _____ Restoration Unit: _____

Time: _____ Temperature and Weather Conditions: _____

North

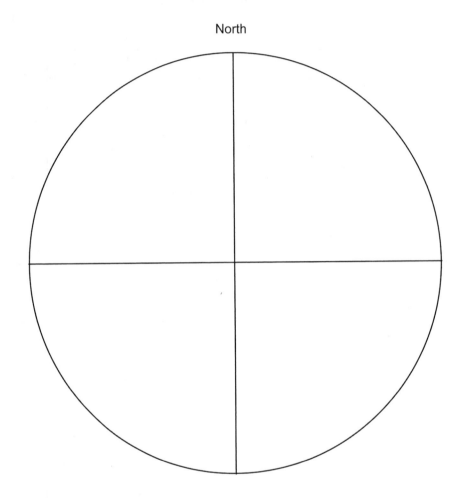

Radius of plot is 50m. Record birds by abbreviations and spell out on back of sheet. Ignore "fly-overs". Record movement in plot by connections abbreviated notations.

From *Restoring Ecological Health to Your Land Workbook* © 2011 Steven I. Apfelbaum and Alan Haney

Data Form 7-1, Restoration Scheduling

IX. Phasing/Scheduling of Tasks:

	Remedial								Maintenance																Who does task	Days of effort required
	Q1	Q2	Q3	Q4	Q1	Q2	Q3	Q4	Q1	Q2	Q3	Q4	Q1	Q2	Q3	Q4	Q1	Q2	Q3	Q4						
	Year 1				Year 2				Year 3				Year 4				Year 5									
1A																										
B																										
C																										
2A																										
B																										
C																										
3A																										
B																										
C																										
4A																										
B																										
C																										
5A																										
B																										
C																										
6A																										
B																										
C																										
7A																										
B																										
C																										
8A																										
B																										
C																										
9A																										
B																										
C																										
10A																										
B																										
C																										

From *Restoring Ecological Health to Your Land Workbook* © 2011 Steven I. Apfelbaum and Alan Haney

Equipment, Safety, and Protection for Restoration Planning

Outdoor clothes and protection. If you are not experienced working outdoors, we recommend that you get local advice on how to be safe and comfortable in the field. Understand your clothing needs; types of footwear and boots to wear; and protection from excessive sun, wind, biting or stinging insects, or venomous animals. If you are going to be walking through brushy areas or in windy and dusty areas, you may need eye protection. Be prepared for changes in the weather. Dress in as many layers as you need for the weather conditions expected. Plan to remove or add clothing, including rain gear, depending on conditions. Last, if you are entering areas with poor communication (e.g., cell phone reception), make sure you have communicated your intentions to at least two coordinated colleagues, and agree ahead on set times you will call in let them know that all is well. Also agree, a priori, that they are to notify local safety or law enforcement personnel if you do not call in at the scheduled time. Even if you are out wandering around your eighty-acre farm, leave a note on the kitchen counter to let others know your intentions.

First-aid. If your first-aid training is dated, we urge that you get the most recent Red Cross first-aid manual and study it carefully. Especially if you will be responsible for others, take a refresher CPR course. Follow the suggestions in the manual for preparing a good first-aid kit, and always keep it with you.

Basic Equipment (needed for most outings)

☐ Clip board—weatherproof to safeguard maps from rain, dew, etc.
☐ Fine point mechanical pencils (2–3 pencils)
☐ Camera with extra batteries or film

☐ Hand-held GPS
☐ Compass (even if hand-held GPS is available)
☐ A 25-foot retractable tape measure
☐ First-aid kit appropriate for setting and conditions
☐ Water bottle with drinking water
☐ Personal toiletries you may need during the day
☐ Cell phone or other communication tools (satellite phone, radio, etc.)
☐ Pocket knife
☐ Bag for collecting plants
☐ Data forms for recording observations and findings

Mapping Equipment

☐ Indelible pens of several colors (2–3 pens)
☐ Transparent Mylar film or frosted acetate film sized to cover the basemap
☐ Plastic laminated copies of the Land-use Cover Codes and Symbols
☐ Plastic laminated mapping codes (see fig. 2.8)
☐ Plastic laminated land-use classifications (see fig. 2.9)

Stream and Soil Inventory

☐ Hand-held soil probe, shovel, or soil auger
☐ Plastic soil-sampling bags
☐ Towel for wiping your hands
☐ Abney level
☐ Level rod for measuring stream terrace relative heights
☐ Fifty-meter tape
☐ Metal pins (used to hold measuring tape)
☐ Yardstick or meterstick
☐ Water bottle for wetting soil
☐ Stopwatch
☐ Soil data form (Profile, Description, and Texture)
☐ Munsel Color Book
☐ County Soil Map and Book that describes soils at your site

Bird Surveys

☐ Binoculars
☐ Bird guide appropriate to your area
☐ Data form for recording observations

Restoration Contracts

We offer four contractual documents to make the process of hiring restoration consultants or contractors, or purchasing seeds or plants, easier.

Disclaimer: In many complex projects, final contract agreements are typically drafted and reviewed by attorneys. Please be advised of this potential need. These documents are intended for educational purposes to help landowners and others streamline the procurement of services and supplies for restoration and are not intended to be final or customized legal agreements.

Contract Grow Agreement. Have nurseries grow seed and plant stock for your project.
Contractor Qualifications. Ensure you select an experienced restoration consultant or contractor.
Design-Build Services Proposal. Explore design-build or turnkey restoration projects with qualified firms.
Proposal for Ecological Restoration Consulting Services. Explore restoration design services with qualified firms or individuals.

Contractor Grow Agreement

Project name: **Delivery #:**   **Order #:**

Client:

Contact name:
Address:
Email:
Phone:
Genetic radius requirements:
Other special requirements:
Client-supplied seed: yes /no
Packaging/shipping instructions:
Product description: Contract growing services for plant species listed and plant numbers, as follows:
Species (common/scientific name) and number of plants of each tray or pot size
(288 128 66 38 2″, 1 qt., $^1/_2$ gal, 1 gal., 5 gal.)

1.
2.
3.

Estimated ready date (based on seed supply date) :

Seed supply date:
Sowing fees for client-supplied seed: $_____per species + $_____per tray
Price per finished plant of each size: $_____
Estimated contract amount: $_____
See attached quote for details

Terms and Conditions: Prices are valid until _____ and do not include shipping or taxes. All items are subject to availability. Terms are net 30. A monthly charge of 1.5% will be charged on all overdue invoices. Items held more than five business days after requested ship date may be subject to storage fees. Contract grow agreements require a nonrefundable sowing fee as a down payment regardless of successful production from client-supplied seed. Sowing fees may be credited to materials purchased in the contract grow. Canceled orders will be charged a 25% restocking fee on the total contract, plus materials for all contract grow agreements. Top growth on plants may be trimmed for shipping unless otherwise requested.

To place order, please sign and return.

Signature and date: _____
Project name and number: _____

Contractor Qualifications

1. Installer qualifications: A contractor with experience in native seeding and native landscape operations with at least 5 years experience. The work shall be in accordance with the standard specifications, except as modified herein.

a. Installer's field supervision: Require contractor to maintain an experienced full-time supervisor on project site who shall be thoroughly familiar with the type and operation of equipment being used when seeding is in progress. Said person shall direct all work performed under this section.

b. Installer's company shall have trained ecologists on staff, who have received Senior Ecologist certifications with the Ecological Society of America, and who are available to provide counsel to the client, to assure quality control over the installers work, and can provide follow-up guidance to the client from to-be-scheduled field visits on the success, short- and long-term maintenance needs of the plantings. Said certified ecologist must have a minimum of 15 years experience and have worked on at least 25 successfully documented projects of comparable size and complexity, in the same or similar ecosystem.

c. Restoring native plant communities is a long-term process. It is imperative that a qualified contractor perform the installation and maintenance of restoration areas for the duration of the maintenance/warranty period.

2. Standards: All materials used during this portion of the work shall meet or exceed applicable federal, state, county, and local laws and regulations.

a. Provide healthy, vigorous, plant materials.

b. Herbaceous perennial plants with severe wilt, badly bruised or damaged stems, or plants with lack of vigorous root systems are not acceptable and may be rejected by the owner.

c. All plant materials shall be free from insects and disease.

d. Species shall be true to their scientific name as specified.

e. All herbaceous perennial plants are to be installed in accordance with the standard specifications shown in the plan, except as modified herein.

3. Materials: The planting contractor shall submit to the owner for approval a complete list of all materials to be used during this portion of the work prior to delivery of any materials to the site.

a. Include complete data on source, amount and quality.

b. This submittal shall in no way be construed as permitting substitution for specific items described in the plans or in these specifications unless approved in writing by the owner.

c. Notify the owner of sources of planting materials 10 days in advance of delivery to site.

4. Provide quality, size, genus, species, and variety of herbaceous perennial plants indicated, complying with applicable requirements in ANSI Z60.1, "American Standard for Nursery Stock."

a. Substitutions will not be permitted without the approval of the owner.

b. If proof is submitted that any herbaceous perennial plant specified is not obtainable, a proposal will be considered for use of nearest equivalent species, size, or variety, with an equitable adjustment to the contract price. Such proof shall be substantiated in writing to the owner.

   c. All aspects of this project have been designed to work together; native plant arrangements and restorations are carefully designed for the planting site conditions as well as species compatibility. Changes to the plans or specifications must be approved in writing by (___contractor's name___) or the owner. (___contractor's name___) is in no way responsible for problems resulting from any changes to the design made by any party without the written permission of (___contractor's name___).

   5. Observation: The owner may inspect herbaceous perennial plants either at place of growth or at site before planting for compliance with requirements for genus, species, variety, size, and quality. The owner retains right to inspect herbaceous perennial plants further for size and condition of root systems, insects, injuries, and latent defects and to reject unsatisfactory or defective material at any time during progress of work.

   6. Remove rejected perennials immediately from the project site.

Retain paragraph below if work of this section is extensive or complex enough to justify a preinstallation conference.

   7. Preinstallation conference: Conduct conference at the project site in order to coordinate equipment movement within planting areas and to avoid soil compaction. Review underground utility location maps and plans. This meeting shall be coordinated by the construction project manager.

   8. Equipment utilized in planting areas shall have low unit pressure ground contact.

Design-Build Services Proposal

For _____ **Project,**

At _____

Introduction

This project includes the following restoration services:

1.
2.
3.

Understandings of Design-Build Service Provider

1. Service provider will design, engineer, specify all services, materials, design elements unless otherwise noted under exclusions spelled out below, for each of the restoration services, required to successfully complete each restoration service listed above.

2. Service provider will obtain all permits, unless otherwise noted below under exclusions spelled out below.

3. Service provide will obtain all materials, unless otherwise noted under exclusions spelled out below, to successfully undertake, install, provide initial installation maintenance to achieve successful establishment and growth. The materials to be provided by the service provider at included in attachment 1 hereto.

4. Service provider will provide the following engineering services in support of the design process and permitting submittals: a_____, b_____, c_____, etc.

5. Service provider will provide the following ecological services in support of the design process and permitting submittals: a_____, b_____, c_____, etc. As an example, the following might be inserted above.

This project involves the restoration and stabilization of approximately 1,500 linear feet of stream reach of the creek immediately upstream of the survey point _____. For this project the service provider will do the following:

- *conduct HSPF hydrologic and hydraulic modeling;*
- *design a series of bioengineering techniques and stream improvements;*

- *assist in obtaining all necessary permits for construction;*
- *remove problematic woody brush;*
- *contract and oversee grading and rock installations, and*
- *construct all bioengineered improvements that include*
  - *live willow staking,*
  - *installation of a 235-foot crib wall with coir fiber rolls and geotextile encapsulated soil lifts,*
  - *brush layering for the crib wall,*
  - *installation of 1,770 native plants, with herbivory (goose) protection, and*
  - *seeding and mulching over 1.5 acres of the streambank and buffer zones.*

6. Service provider will also provide homeowner communication and educational services to ensure acceptance and cooperation from the residents adjacent to the said project. At the conclusion of the project, service provider will produce and deliver a homeowner's guide to the understanding and care of the restoration project.

Project Approach

The project approach is a most important part of the design-build contract and should include the following type of content, at an appropriate level of detail, and in decisive language for clarity and to serve as a binding contract between parties.

- How will this be designed and built?
- Who from the service provider company will be the lead with whom communications and responsibilities for success reside?
- What is the schedule for each task?
- What are the start and completion dates?
- What arrangements should be stated regarding delays, including those caused by the service provider, weather, or other subcontractors or suppliers?
- Who is in charge of construction access, special details on equipment and materials staging locations and needs, including irrigation needs for plant stock?
- How should cleanup, erosion control, waste materials, and debris/garbage management be addressed?

The following scope of services describes service provider's proposed work plan for complete design/build services.

Exclusions

1. If, during the design phase, we conclude that hydraulic forces require more traditional, hard engineered strategies, we will notify the owner of our recommendation and provide preliminary cost estimates for anticipated solutions.

2. Plant and seed availability may change from time to time and service provider will provide owner with substitutions as they become known.

3. Etc.

4. Etc.

The following scope of services is divided into three phases: Phase I—Design; Phase II—Construction; and Phase III—Three-year Management. At this time, we have not attempted to provide a schedule with timing of these design-build activities, due to the uncertainty of budgeting and the timing of the _____ grant funding request and fund availability. However, if the owner is prepared to begin the project this year, service provider is prepared to begin the design phase immediately, with construction tasks anticipated to commence _____.

Scope of Services

Phase I: Planning

Task 1, *Kickoff meeting*: _____ staff will meet with you and your representatives to obtain additional details related to the proposed work plan, relationships with property owners, expectations, and to finalize a scope of service and schedule.
Deliverables: Meeting minutes
Fee estimate

Task 2, *Base information*: service provider staff will use a color, digital, orthorectified aerial photograph as a basemap for all subsequent layers of information including topography, inlets and outlets, roads, floodplain, and property boundaries. All layers available digitally will be superimposed on the basemap using AutoCad. We understand that the owner will provide service provider with existing property boundary and engineering data on this stretch of waterway.
Deliverables:
CAD drawings
Fee estimate

Task 3, *Field assessment and survey*: Service provider staff will develop a topographic drawing identifying the centerline of the channel, thalwag, toe of slope, and top of slope. The survey shall be based on at least 10 cross sections. All trees greater than 4 inches dbh will be tagged and surveyed. All bridges, pipes, inlets, and outlets will be surveyed.
Deliverables:
Digital topographic drawing
Fee estimate

Task 4, *H & H modeling*: Service provider staff will use hydrologic model PondPack to develop runoff flows for design storm events from the watershed and use hydraulic model HEC RAS to define depth of flow, and flow velocities in the study area.
Deliverables:
Report summarizing results
Fee estimate

Task 5, *Engineering*: Service provider staff will prepare an engineering drawing with sufficient detail to permit and implement. Plans shall be stamped by a P.E. licensed in _____ as required by the _____.

Deliverables:

Report summarizing results

Fee estimate

Task 6, *Permitting*: Service provider will obtain necessary permits and approvals from the US Army Corps of Engineers, _____ Storm Water Management Commission and other as follows:

_____, _____, _____.

Deliverables:

Permit applications

Fee estimate

Phase II Implementation

Service provider shall construct the project per the plans. Anticipated tasks are described below: insert Excel spreadsheet

Phase III Maintenance and Management

Service provider shall be responsible for three years of maintenance as follows:

Fee estimate

Year 1:   Herbiciding
          Mowing

Year 2:   Herbiciding
          Mowing

Year 3:   Herbiciding
          Mowing
          Prescribed burn

Proposal for Ecological Restoration Consulting Services

[Letterhead] (Note: Use the appropriate letterhead for your office and/or department.)

[Date]
[Client Name]
[Address]
RE: Proposal for Ecological Restoration Consulting Services for _____ project _____, _____ township, _____ county, _____ city or village, state, for _____ client_____ (Project# XX-XXXX)

Dear _____,

Thank you for the opportunity to provide you with this proposal for ecological consulting services related to _____. Attached please find our scope of services and fees based on our understanding of your request for our services.

    We are confident you will find that we provide exceptional expertise, service, and value. We look forward to beginning work with you. Please call if you have any questions about the attached proposal and supporting documents.

    Sincerely,
    [Name and title]
    Attachments

Understanding of Assignment

A consultant should repeat back to the client their understanding of the client's needs and the deliverables requested. Place the project into context with its surroundings, identify the main problems to be addressed/solved, and other pieces of background information to justify the scope of work provided below. Begin paragraphs with "We understand that . . ." or "It is our understanding that. . . ." Or "You informed us of . . ." State what services the client has asked consultant to prepare a proposal for. Describe or at least state consultants' qualifications to perform the work. Provide a more formal statement of qualifications for particular expertise if you feel it is necessary or if the client has asked for particular qualifications and team member information. End with "Below please find our scope of work" or "Our proposed scope of work is outlined below."

Project Approach

Add, if necessary, especially if our approach is different than what the client has asked for.

Scope of Work

Task 1: _____. (Describe in paragraph form.)
Deliverable: _____
Schedule:_____
Estimated fee: $_____

Task 2: _____. (Describe in paragraph form.)
Deliverable: _____
Schedule:_____
Estimated fee: $_____
Total estimated fee: $_____ or
Total estimated fee not to exceed: $_____ or
Lump sum fee: $_____

(Identify fees per task if the proposal is to be a time and materials estimated fee contract. Identify a lump sum fee if not.)

Schedule

Consultant will work out a schedule of deliverables with the client.
or
Consultant will perform work according to the schedule identified above.

Payment

(Include all that apply.)

Estimated fees: Fees noted "estimated" are estimates only. Consultant will bill on a time plus expenses basis according to the attached fee schedule. Total billings may be higher or lower than that estimated.

Lump sum: Fees noted as "lump sum" will be billed monthly based on an estimated percentage of completeness.

Estimated fees, not to exceed: Consultant will bill on a time plus expenses basis according to the attached fee schedule. Total billings shall not exceed the total maximum billable amount of the contract unless explicit approval has been given by the client for additional services and related costs.

Additional Services

Consultant will provide additional services, above and beyond the scope presented above, with explicit approval from the client. Fees will be based on the fee schedule attached to this contract.

Special Conditions (Delete if unnecessary.)

The following special conditions apply to this scope of work:

1.
2.
3.

(If necessary list out what your proposal does *not* include, or your assumptions used in preparing the scope and fees, or what you must get from the client or others in order to do the work as proposed.)

Attachments

Consulting Contract Short Form
Fee Schedule for Consultants and Support Services
General Terms and Conditions

Valid Period

This contract is valid for a period of one year from the date appearing at the top of page one.

Approval

In signing the attached Consulting Contract Short Form each party agrees to abide by all terms and conditions presented in this document and the defined attachments. Please sign and return *two* original copies to consultant. One original with signatures from both parties will be sent back to you.
We thank you the opportunity to provide these services on your project.

_____

Signature of consultant
Title of consultant

_____

Acceptance signature of consultant proposal — signature of client
Date _____

Additional Resources for Readers

The Restoring Ecological Health to Your Land Workbook and its companion book, *Restoring Ecological Health to Your Land*, are focused on the essential information to help landowners, public agency personnel, stewardship volunteers, and others to begin and successfully implement restorations. However, we know from our own restoration projects that at any point in the process, challenges may arise where you might benefit from more help than has been provided in this or any other book.

Time and again in our restoration careers we wished we had access to the authors of books, or other experts, as even the best " how-to-do-it" books left us with some puzzling questions. We hope to help you through such challenges and offer special arrangements for you to lean on us. We offer the following resources to help you get started on your restoration project and to successfully carry it through to completion.

Website: www.restoringyourland.com

We have created a website where you can download the data forms also found in the book and to give you access to interactive data forms and mapping routines to help you develop your restoration plans. To gain access to these materials, you must correctly answer a qualifying question that will be posted on the website.

Free Telephone Consultation

If you want further help from us, go to the restoration book series website, www.restoringyourland.com, and follow the instructions to complete and send us the restoration project application. The application will provide us with basic details about you and your project. Once we review the application, we will send you an email to arrange a one-hour, free telephone consultation. During this call, we will attempt to help you address the hurdles you may find challenging.

Webinar Presentations

If you are students or professors using the book in your classroom, or part of a book club, we will be delighted to use the one-hour period to present a telephone webinar to explain the book and the restoration process and to address questions you may have. Simply explain this educational intent in completing the application.

Workshops

If you represent a group or are one of multiple landowners working together to undertake a restoration, after the free one-hour telephone consultation, we can work with you to develop and deliver a workshop for invited attendees. For basic expenses, we will walk participants through the restoration process, on the restoration project site. We have completed a number of such events throughout the United States and have learned that such workshops can bring neighbors together to design and commit to achieving restoration objectives beyond individual landowners, can introduce others to the project, and can be a very efficient way to jump-start a broad restoration program.

Budget Preparation

We will work with you to prepare a budget to cover basic expenses, as a part of the project design process, if desired, after the one-hour of free consultation.

We are committed to helping your restoration project succeed.

THE SCIENCE AND PRACTICE
OF ECOLOGICAL RESTORATION

Wildlife Restoration: Techniques for Habitat Analysis and Animal Monitoring, by Michael L. Morrison

Ecological Restoration of Southwestern Ponderosa Pine Forests, edited by Peter Friederici, Ecological Restoration Institute at Northern Arizona University

Ex Situ Plant Conservation: Supporting Species Survival in the Wild, edited by Edward O. Guerrant Jr., Kayri Havens, and Mike Maunder

Great Basin Riparian Ecosystems: Ecology, Management, and Restoration, edited by Jeanne C. Chambers and Jerry R. Miller

Assembly Rules and Restoration Ecology: Bridging the Gap Between Theory and Practice, edited by Vicky M. Temperton, Richard J. Hobbs, Tim Nuttle, and Stefan Halle

The Tallgrass Restoration Handbook: For Prairies, Savannas, and Woodlands, edited by Stephen Packard and Cornelia F. Mutel

The Historical Ecology Handbook: A Restorationist's Guide to Reference Ecosystems, edited by Dave Egan and Evelyn A. Howell

Foundations of Restoration Ecology, edited by Donald A. Falk, Margaret A. Palmer, and Joy B. Zedler

Restoring the Pacific Northwest: The Art and Science of Ecological Restoration in Cascadia, edited by Dean Apostol and Marcia Sinclair

A Guide for Desert and Dryland Restoration: New Hope for Arid Lands, by David A. Bainbridge